SOTHIS PRESS

Sothis Press aims to completely reimagine the public domain book with an emphasis on universal access to captivatingly designed print editions. We believe that technology will enable every single book ever published to be perpetually accessible in beautiful print form. Sothis is beauty in ubiquity.

www.sothis.press
1 Waterhouse Square, London, ECIN 2ST

Geometrical Optics

With Diagram

Archibald Stanley Percival

SOTHIS PRESS

ISBN 978-1-0365-1631-4

Geometrical Optics is in
the public domain.

GEOMETRICAL OPTICS

BY

ARCHIBALD STANLEY PERCIVAL

M.A., M.B., B.C. CANTAB.

SENIOR SURGEON NORTHUMBERLAND AND DURHAM EYE INFIRMARY

AUTHOR OF "OPTICS," "PRACTICAL INTEGRATION," "PRESCRIBING OF SPECTACLES"

WITH DIAGRAMS

LONGMANS, GREEN, AND CO.

39 PATERNOSTER ROW, LONDON

NEW YORK, BOMBAY, AND CALCUTTA

1913

PREFACE

THIS book is primarily intended for medical students as a text-book on the subject of Geometrical Optics for their preliminary scientific examinations, though it practically contains all the Optics required by an ophthalmic surgeon. It is hoped that it will also prove of service to students of physics, as some knowledge of the subject is indispensable if the laboratory experiments are to be understood.

It requires prolonged and deep study to form any vivid conception of the now accepted theory of light, and in all elementary books the form in which the undulatory theory is presented is so crude that it is both untrue and useless. The subject of Physical Optics therefore has been avoided entirely; indeed, I am convinced that no thorough elementary knowledge of that intricate subject can be obtained in the short space of time allotted to the student for studying Optics.

As an introduction to mathematical analysis the subject of Geometrical Optics has no equal, for it insists on the importance of paying due attention to the meaning of algebraic signs, and it is also an easy introduction to several somewhat difficult mathematical conceptions. For instance, the vectorial significance of the line BA being considered as equal to $-AB$, or AB taken in the reverse direction, opens up a new vista to the student of Euclid and elementary geometrical methods: equally novel is the conception of a virtual image. At the same time every student can verify for himself the results of his calculations so simply by experiment that it will convince him of the reality of the analytical methods employed.

I have embodied in the text all that can be reasonably demanded in any preliminary examination in science, while the Appendix contains matter that will be of service subsequently in a professional career. The starred paragraphs may be omitted at a first reading, as a knowledge of them would be rarely required in the examination-room. The subject of Cardinal Points is always neglected in the elementary books; it has here been treated, I venture to think, in a much simpler manner than is customary in the solution of the problem; for the ingenious graphic solution on p. 106 I am indebted to Professor Sampson. The subject is of first importance in understanding the optical properties of the eye, and it is of the greatest value in dealing simply and readily with many otherwise difficult questions.

The reader will find a list of all the important formulæ for ready reference at the end of the book, and he will notice that some of them are of universal application, and that by a simple transformation, formulæ for refraction can be converted into the corresponding formulæ for reflection. The proofs while preserving their rigid character are made as simple as possible, and the utmost care has been taken to include only what is of practical application, excluding all that is of merely academic interest.

ARCHIBALD STANLEY PERCIVAL.

17, Claremont Place,
Newcastle-upon-Tyne,
Dec. 1912.

CONTENTS

CHAPTER I

ILLUMINATION—PINHOLES—SHADOWS

WE shall here confine ourselves to the study of some of the simplest properties and laws of Light. When we say that we see the sun or a tree we merely mean that we see the light that comes from them; the sun, of course, sends out light of its own, whereas the tree passes on or reflects the light that it receives from something else. It is obvious that in neither case do we see the thing itself; we are only conscious of a certain sense-impression derived from it. The nature of this sense-impression and the way in which it is developed from a physical stimulation of the retina, are problems that are still engaging the attention of physiologists and psychologists; with questions such as these the science of Optics does not deal.

We shall commence our study by the consideration of the following laws:—

LAW 1. In a homogeneous medium light is propagated in every direction in straight lines.

LAW 2. The intensity of illumination varies inversely as the square of the distance from the source of light, and it is greatest when the angle of incidence is 0.

LAW 3. When the incident rays of light are parallel, the intensity of illumination varies as the cosine of the angle of incidence.

Our experience of shadows confirms the truth of Law 1. A luminous point in space sends out light in all directions, illuminating a rapidly enlarging sphere. It is the radii of this sphere that are called rays of light; they are simply the directions in which the light is travelling, and when this direction is altered by reflection or refraction the rays considered are bent at an angle. But for either reflection or refraction to occur a heterogeneous medium must be encountered; so that we conclude that in a homogeneous medium light is propagated in every direction in straight lines.

Experiments with diffraction gratings show us that Law 1 is not strictly true, and, had we to treat of the phenomena of Diffraction and Polarization, we should have to explain in detail the electromagnetic theory of light, or at any rate to give an account of some form of wave theory. As these subjects do not now concern us, we may regard the directions in which light travels as straight lines or rays, and in this way the elementary study of optics will be much simplified.

The second and third laws will require some careful explanation and consideration before they will be accepted by the reader.

Illumination.—Let S be a source of light, and consider the light that is being propagated from S in the direction of the screens AK and BL (Fig. 1). It will be noticed that,

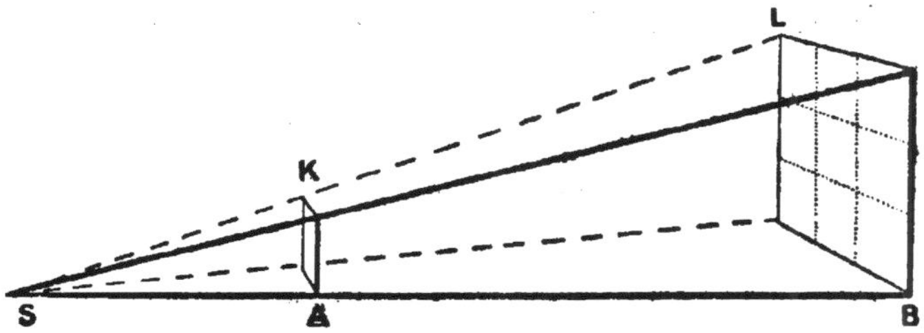

Fig. 1.

since light travels in straight lines, the amount of light (Q) that falls upon the square AK is the same as that which would illuminate the larger square BL if AK were removed. Now, in the diagram SB is made equal to 3 SA, so the area of the square BL is nine times the area of AK. Hence the illumination per unit area, or the *intensity of illumination*, of BL is $\frac{1}{9}$ of that of AK.

Indeed,
$$\frac{Q}{BL} : \frac{Q}{AK} = 1 : 9.$$

It is obvious, then, that the intensity of illumination varies inversely as the square of the distance from the source of light.

Nearly all photometers are based upon this law; we will describe that designed by Rumford. Suppose that we wish to compare the intensities, l and L, of illumination of two lamps; these are placed at distances d and D from a white screen so that two *sharp* shadows of a vertical rod placed a few inches in front of the screen are thrown upon it; each shadow is then illuminated by only one of the lamps. Thus if L, the standard light, be regarded as 1, the other lamp is moved backwards or forwards until the position d is found in which both shadows appear equally dark. Each lamp is then sending an equal amount of light to the screen, and the relative illuminating power is given by the ratio of the squares of their respective distances (d^2 and D^2) from the screen. The light that the two shadows receive are respectively $\frac{l}{d^2}$ and $\frac{L}{D^2}$; consequently when these are equal, $l : L = d^2 : D^2$. Thus if D the distance of the standard light be 2 feet, while $d = 4$ feet,

$$l : 1 = d^2 : D^2 = 16 : 4 = 4 : 1$$

or l is four times the intensity of the standard light.

It is commonly said that four candles at a distance of 2 feet give the same illumination as one candle at a distance of 1 foot; this, although true, could not be satisfactorily proved by a photometer which measures only the intensity of illumination from a point source of light.

It may be noted that the accuracy with which such tests can be made depends upon the "Light Sense" of the observer, so that if the values of l and L be known, this photometer may be used as a test of the observer's "Light Difference."

On referring to Fig. 1, it will be noticed that both AK

and BL are at right angles to SB. Now, the angle of incidence (ϕ) is the angle which a line at right angles to the surface of the screen makes with the incident ray; in this case, therefore, the angle of incidence is 0. If the screen were inclined either forwards or backwards, turning round its base line, it is clear that the intensity of illumination per unit area of the screen would be diminished, for in either case some of the cone of light that is represented would not fall upon the tilted screen. The diminution of illumination approximately follows the cosine law (see below) when the incident cone of light is not too divergent.

If the illuminant were a beam of parallel rays, or were at an infinite distance as compared with the distance AB, the law of inverse squares ceases to have any intelligible meaning; for then both SA and SB are infinite, and the illumination of a surface exposed to such a light is constant whatever its distance. For instance, the light from a morning sun on a screen is practically the same as the light on a similar screen placed at a mile's distance to the west.

If I be the intensity of illumination a screen receives from a beam of parallel rays when the angle of incidence is 0, it is clear that when the screen is tilted so that the angle of incidence is ϕ, the illumination is $I \cos \phi$.

Apparent Brightness.—The distinction between brightness and illumination is hardly made clear enough in the books. Illumination refers to the physical condition of the object when illuminated, whereas brightness refers to the resulting sense-impression produced in the observer. It will clearly depend upon his physiological condition; but, neglecting this for the moment, let us consider how an illuminated screen at BL will appear to an observer at S.

The brightness of an object is naturally measured by the amount of light it sends to the eye per unit area of its apparent size; in other words, the brightness (B) of an object is directly proportional to the quantity of light (Q) that it sends to the pupil, and inversely proportional to the apparent area (A) of the surface observed, or $B = \frac{Q}{A}$.

Since, however, both Q and A are each of them inversely proportional to the square of the distance of the luminous object, the apparent brightness is independent both of its size and of its distance, presuming of course that the medium is clear and no adventitious absorption occurs in it. If the physiological conditions be the same, the apparent brightness of an object merely depends upon the intrinsic luminosity of the object. A red-hot iron ball is indistinguishable from a circular disc of iron at the same temperature, showing that the brightness is independent of the inclination of the periphery of the ball to the line of sight. The sun and moon may both be regarded as approximately spherical, yet both appear to the naked eye as flat discs of uniform brightness. There are, however, several physiological conditions that greatly affect the apparent brightness of luminous bodies.

(1) *The size of the pupil.* The brightness (B) must vary as the area of the pupil. The brightness of an object will be much diminished by viewing it through a small pinhole, for in that case less light enters the eye, as the size of the pupil is virtually diminished.

(2) *The condition of retinal adaptation.* Prolonged stay in a dark room will much increase the apparent brightness beyond what is due to the consequent dilatation of the pupil.

When high powers of a microscope or a telescope are used, the brightness of the image is much diminished, as then only part of the pupil is filled with light. This is practically equivalent to lessening the area of the pupil (1). It is impossible by any optical arrangement to obtain an image whose brightest part shall exceed the brightest part of the object. If allowance be made for the loss of a certain amount (about 15 %) of light by reflection and imperfect transparency of the lenses, the brightness of the image is equal to the brightness of the object.

There is one case in which this law does not hold good, and this also depends upon physiological reasons. If the object be extremely small, it may yet if excessively bright succeed in stimulating a retinal cone, and so cause a visual impression of a tiny point of light, even although its image does not cover the whole surface of the cone. If such an object be magnified until its image covers the cone, there will be no increase of its apparent size although an increased amount of light will be entering the eye. When, for instance, stars are observed through a telescope, their apparent size is not increased, for their image does not extend beyond one retinal cone, but all the light that falls on the object-glass may by a suitable eyepiece be concentrated on the observer's pupil, except that lost by transmission through the lenses. If, then, a denote the fraction of the incident light that is transmitted through the telescope (about 0·85), and O denote the area of the object-glass, and e that of the pupil, the increase of brightness will be $a\frac{O}{e}$. Or if o and p be the diameters of the object-glass and the pupil respectively, the increase of brightness will be $a\frac{o^2}{p^2}$, for the areas of circles are proportional to the squares of their diameters. If the pupil be regarded as of unit diameter we get the expression ao^2. This is what astronomers call the "space penetrating power" of a telescope, that is its power of rendering very small stars visible.

Precisely the same conditions obtain with the ultramicroscope, which is a device for bringing into view fine granules that are smaller than the limit of the resolving power of the instrument. By means of dark-ground illumination these fine particles are relatively so brightly lighted that they succeed in stimulating a retinal cone even though their image does not cover its whole surface. Resolution does not occur; whatever its real shape the granule will appear as a circular point of light, so that all one can really say is that something very small is present.

Visual Angle.—On referring again to Fig. 1 it will be seen that if the observer's eye were situated at S the small square AK would completely obliterate the larger square BL, as they both subtend the same angle at S. Such an angle is called the visual angle, and it is obvious that if one does not know the distance of an object one can form no real idea of its size. The expression "apparent size" is quite illusory; it conveys no clear meaning. For instance, some will say that the sun's apparent size is that of a dinner plate, others that of a saucer, but it will be found on trial that a threepenny bit at a distance of 6 feet will completely obliterate it, yet few would assert that its apparent size was that of a threepenny bit even when held at arm's length. The importance of the visual angle will appear when we deal with magnification.

Pinholes.—If a pinhole be made in a card, and this be held between a candle and a screen, an inverted image of the candle flame will be formed upon the screen. The nearer the screen is brought to the pinhole the smaller and sharper will be the image. If the candle be brought nearer the pinhole, the image will be larger but less sharp. Finally, if the hole in the card be made larger the image will appear brighter, but its definition will be again diminished.

The explanation is simple, as will appear from an examination of Fig. 2. Every point of the candle flame is sending out light in all directions, and all that falls upon the card is intercepted by it, so that its shadow is thrown upon the screen. From each point of the candle flame, however, there will be one tiny cone of light that will make its way through the pinhole. On the screen the section of this cone will appear as a small patch of light. Thus the point A of the flame will be represented by the patch *a*, and the point B by the patch *b* on the screen. Clearly the height *ab* of this inverted image will be proportional to the distance of the screen from the pinhole; if AB be brought nearer the card each pencil will form a more divergent cone, or if the hole be made larger the same result will occur; hence in both these cases the patches *a* and *b* will be larger and the definition will be impaired.

If care be taken that these patches of light are not too large, an image sufficiently sharp for photographic purposes may be obtained. For some purposes a pinhole camera made out of a preserved meat tin may prove a more satisfactory instrument than one provided with a lens. The disadvantage of a pinhole camera is of course the feeble illumination of the image, which makes a very long exposure necessary.

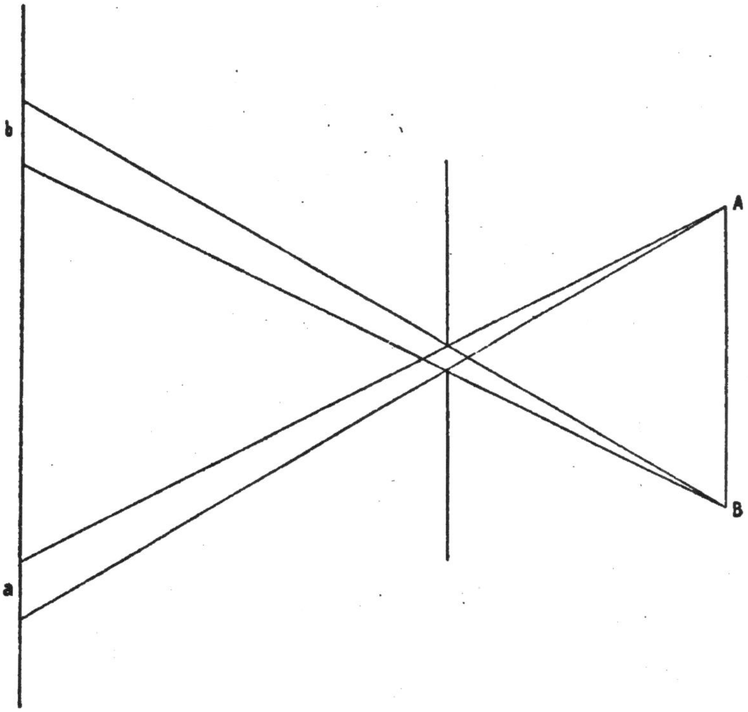

FIG. 2.

The best size of pinholes for photographic work, according to Abney, is determined by the approximate formula—

$$p = 0{\cdot}008\sqrt{d}$$

Thus, if the distance between the pinhole and the plate be 9 ins., the diameter of the pinhole should be 0·008 × 3 = 0·024 in.

Shadows.—When the source of illumination may be regarded as a point, it is clear that an opaque obstacle like AK in Fig. 1 will be lighted only on its proximal surface, while the prolongation of the cone of light indicates the cone of shadow cast by AK.

If, however, the source of light be a luminous body possessing innumerable points, the object will be illuminated by innumerable cones of light, and we must imagine a shadow for each of them behind the object. The space behind the object which is common to all of these shadow cones will represent the area of total shadow, or *umbra*. There will, however, be a space outside this which is only in shadow as regards part of the luminous body while it receives light from another part of it, and is consequently partially illuminated. This is the area of part-shadow, or *penumbra*. In the adjoining illustration (Fig. 3) two opaque bodies are represented,

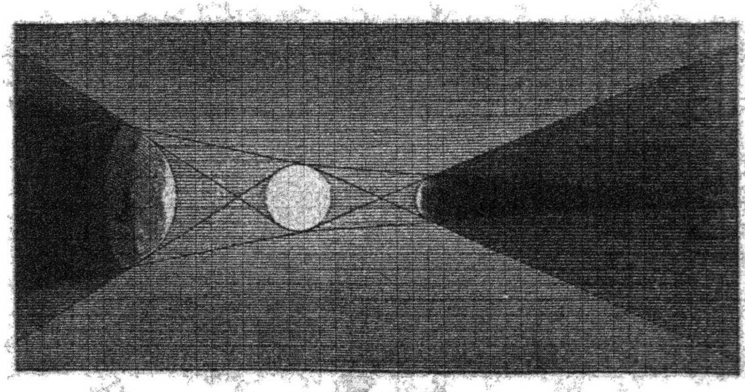

FIG. 3.

one being smaller and the other larger than the luminous body which is placed between them. In each case for the sake of clearness the limits between umbra and penumbra have been more sharply defined than they should be. When a total eclipse of the sun occurs, the moon is so situated that it intercepts all the light coming from the sun towards the earth; the earth is then in the *umbra:* when a partial eclipse occurs, the earth is in the *penumbra*.

QUESTIONS.

(1) What is the height of a tower that casts a shadow 52 feet 6 inches in length on the ground, the shadow of a stick 3 feet high being at the same time 3 feet 6 inches long?

(2) A pinhole camera, the length of which is 7 inches, forms an inverted image 4 inches high of a house that is in reality 40 feet high. What is the distance of the camera from the house?

(3) A gas lamp distant 5 feet and an electric light distant 150 feet throw on an opposite wall two shadows of a neighbouring post. If these two shadows are of equal intensity, what is the relative illuminating power of the lights?

(4) If the electric light in (3) were raised vertically to such a height that its distance from the wall were 300 feet, what would be approximately the relative intensities of the shadows?

CHAPTER II

REFLECTION AT PLANE SURFACES

WE must now consider the manner in which light is reflected by polished surfaces. The following two laws when properly understood explain every possible case of reflection:—

LAW 1. The reflected ray lies in the plane of incidence.

LAW 2. The angles which the normal (at the point of incidence) makes with the incident and with the reflected ray are numerically equal, but they are of opposite sign.

These laws hold good whether the surface be plane or curved; in the latter case we have only to draw a tangent at the point of incidence, and consider the ray reflected at this plane. The meaning of the technical terms and the sign of an angle will be explained in the next paragraph.

Reflection at Plane Surfaces.—Let AB represent a plane reflecting surface (Fig. 4), and let S be a luminous point sending out light in all directions. Now, if SN be one of these directions, SN represents an incident ray, N the point of incidence, and NY the normal at that point. Now, the plane of incidence is that plane that contains both the incident ray and the normal, so the plane of the paper is the plane of incidence. According to the laws just stated NQ is the corresponding reflected ray, for the angle of incidence YNS is numerically equal to the angle of refraction YNQ, which lies in the plane of the paper, and is of opposite sign to YNS. The angle YNS is measured by the rotation at N of a line in the direction NY to the position NS, *i.e.* in the direction of the hands of a clock, whereas YNQ is measured in the counter-clockwise direction. This is what is meant by the sign of an angle, so that if YNQ be denoted by ϕ' and YNS by ϕ, we have in reflection $\phi' = -\phi$.

If, then, there be an eye in the neighbourhood of Q it will receive light coming towards it in the direction NQ; but we have not yet found from which point in this line it will appear to have come. To do this we shall have to take another incident ray, SM say, and discover where the corresponding reflected ray intersects the previous one.

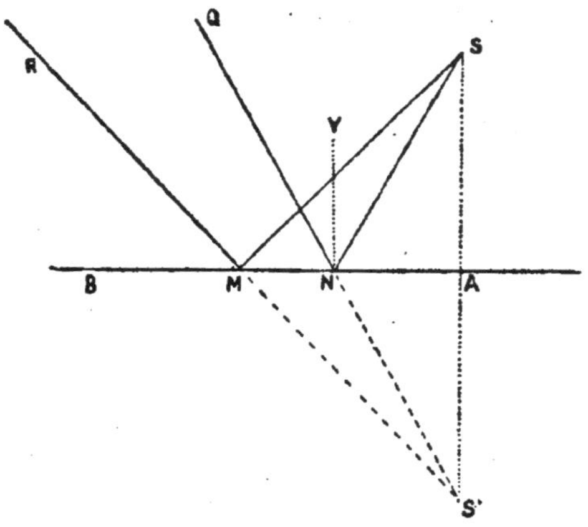

Fig. 4.

It is clear that as the incident and reflected rays make equal angles with the normal they must make equal angles with the mirror; so we made the angle BMR numerically equal to AMS, and produce the reflected ray RM to meet QN produced in S'. Then S' is the point from which the light will appear to have come.

Now, it is clear that in the triangles SMN and S'MN we have the base MN common, and the angles at the base equal, so the triangles must be equal; and if we join SS', we see that the two sides SM, MA are equal to the two sides S'M, MA, and the included angle SMA is equal to S'MA; so S'A is equal to SA and the line SS' is at right angles to AB.

Similarly, it may be shown that any other ray in the same

plane will be reflected in such a direction that when produced backwards it will meet SS′ in S′. Now, if we suppose the paper to be revolved about an axis SS′, the figure will represent the course of incident and reflected rays in every plane. Hence all the rays that fall upon the mirror from S, whatever the plane of their incidence, will be so reflected that the prolongations of these reflected rays will intersect at the point S′, so that an image of S will be formed at S′.

It will be noted that this image has no real existence; the reflected rays do not really come from the back of the mirror, they only appear to come from the point S′: such an image is called a *virtual* image. Subsequently we shall deal with *real* images, and the distinction between them is simply this: *Real images are formed by the intersections of the reflected (or refracted) rays themselves; virtual images are only formed by the intersections of their prolongations.*

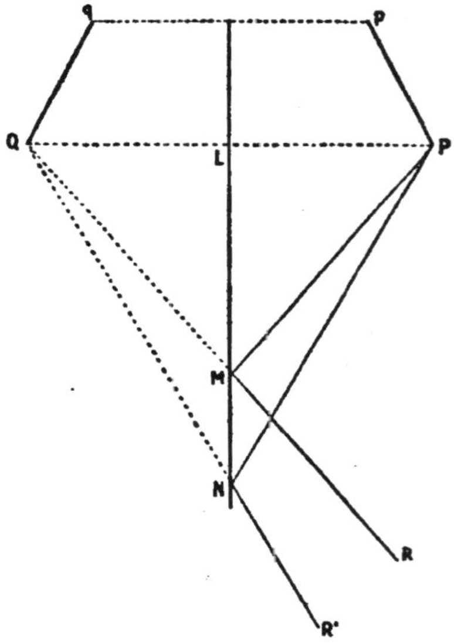

FIG. 5.

Construction of the Image of an Object.—Suppose that it is required to construct the image of the object P*p* as seen in the mirror NM (Fig. 5); all we have to do is to draw lines perpendicular to the mirror (or its prolongation) from P and *p* to Q and *q*, so that Q and *q* are as far behind the mirror as P and *p* are in front of it; then Q*q* is the virtual image of P*p*. If an eye be situated in the neighbourhood of R, and we wish to show the actual course of the light that reaches the eye from P, we join RQ, cutting the mirror at M, and then we join MP. The light that forms the image Q for an eye at R reaches it by the path PM and MR. Another eye at R′ will see the same image Q by means of light that travels in the direction PN and NR′.

It will be noted that the image Qq is similar and equal to the object Pp, for they both are similarly situated *with respect to the mirror.* Since, however, the object and the image face each other, the observer gets a view of that side of the object that he could not see without turning it round. Hence he thinks the object has been turned round so that its right and left sides have been interchanged. For this reason the image of a printed letter will appear "perverted," that is, it will appear upright, but it will resemble the type from which it was printed.

Deviation produced by Rotation of Mirror.—If the mirror AB (Fig. 6) be slowly rotated into the position *ab*,

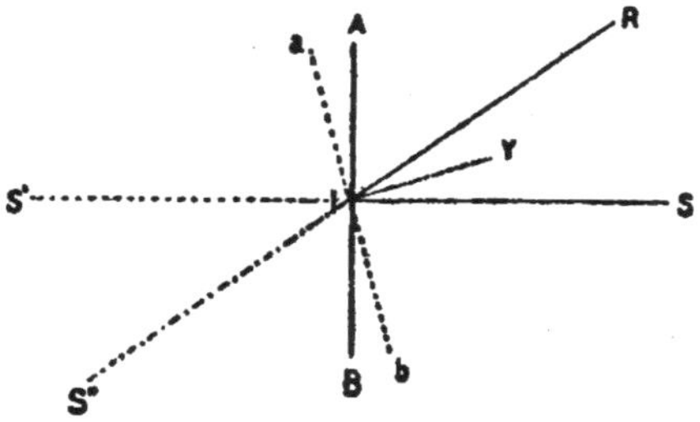

Fig. 6.

the image S′ will appear to rotate through an angle twice as great in the same direction, so that finally it will appear to have moved to S″. The explanation is obvious. If S be the object, when the mirror is in the position AB its image will be at S′; on rotating the mirror through an angle θ its normal will also rotate through the same angle, so that SIY $= \theta$, but then reflection will occur in the direction IR, and the image of S will appear at S″. Clearly, S′IS″ = SIR $= 2\theta$.

There are several important applications of this principle in daily use; as examples we may mention the sextant and the mirror galvanometer, but we will describe in greater detail the laryngoscope. When this instrument is used, the mirror, which is inclined at an angle of 45°, is placed at the back of the pharynx; when properly placed an image of the larynx is seen with its axis horizontal. Since the image and object are similarly situated with respect to the mirror and not to the observer, the anterior parts of the larynx (epiglottis, etc.) are represented in the upper part of the image, while the posterior structures (arytenoids, etc.) occupy the lower portion of the image.

Repeated Reflection at Inclined Mirrors.—When an object is placed between two plane mirrors inclined at an angle, a limited number of images may be seen by an observer in a

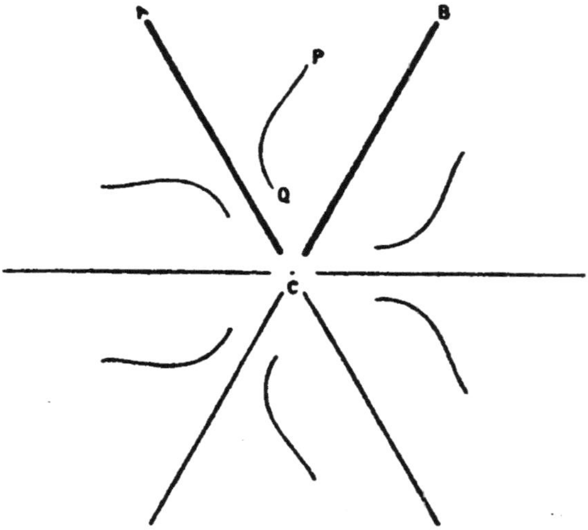

FIG. 7.

suitable position. If a be the angle between the mirrors, and if $\frac{180^\circ}{a} = n$, the number of images seen by an eye in a suitable position will be $2n-1$, if n be an integer. Moreover, the object together with its images then forms a perfectly symmetrical figure with respect to the reflecting surfaces. Fig. 7 represents two mirrors, AC and BC, inclined at an angle of 60° with an object PQ between them. As $2n - 1$ in this case is 5, 5 images are seen, and the symmetry of the figure is evident. This is the principle of the kaleidoscope.

Suppose that $a = 40°$, then $n = \frac{180°}{40°}$ or $4\frac{1}{2}$; in such a case it will depend entirely on the position of the observer whether 4 or 5 images are seen.

The position of the images is easily found by the construction now familiar to the reader. Let AC, BC (Fig. 8) be two plane mirrors inclined at an angle of 90°, and let p denote the position of an object between them. From p draw perpendiculars to each mirror, and produce them to p_1^a and p_1 so that p_1^a and p_1^b are as far behind the

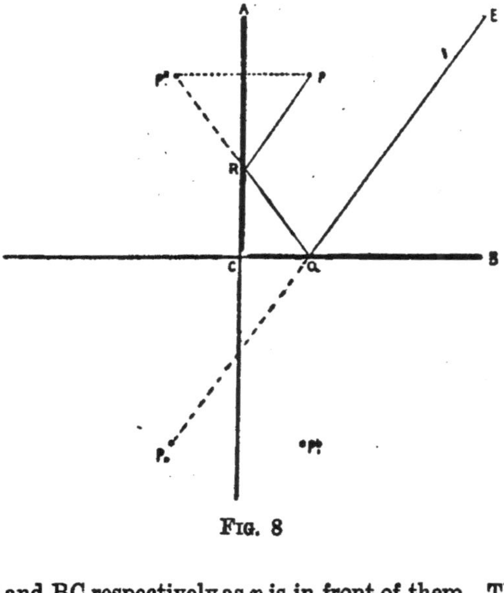

FIG. 8

mirrors AC and BC respectively as p is in front of them. Then p_1^a and p_1^b are two of the images. To find the position of the third image p_{11}, we must produce AC and BC and consider these prolongations as mirrors and the images we have already found as objects. That is to say, we make p_{11} either at an equal distance behind AC produced to that of p_1^b in front of it, or we make p_{11} an equal distance behind BC produced to that of p_1^a in front of it. It will be noticed that whichever construction is used, p_{11} occupies the same place. To find the path of light from p_{11} to an eye at E, join p_{11}E, cutting BC in Q; join p_1^aQ, cutting AC in R; join pR. Then the path of the light from p is pR, RQ, QE. If E had been close to AC, on joining p_{11}E the mirror AC instead of BC would have been cut, so that in such a case the last reflection would have been from the mirror AC, and p_{11}

must be joined to the intersection on AC. These constructions are of little importance except for examination purposes.

When the mirrors are parallel to each other, the angle between them is 0, and since $\frac{180^\circ}{0^\circ} = \infty$, an infinite number of images might be seen if certain physical conditions did not prevent their observation.

QUESTIONS.

(1) Two parallel mirrors face each other at a distance of 8 feet, and a small object is placed between them at a distance of 1 foot from one of them. Calculate the distances from this mirror of the three nearest images that are seen in it.

(2) Two mirrors, AC and BC, are inclined at an angle of 45°; an object, P, is so placed that ACP = 15°. How many images will an observer halfway between the mirrors see? Trace the path of light that gives rise to the second image $p_{11}{}^b$ seen in BC.

(3) If in the last example BC were rotated so that ACB = 135°, how many images would an observer (E) halfway between the mirrors see? In what position would he see more?

CHAPTER III

REFLECTION AT A SPHERICAL SURFACE

REFLECTION at irregularly curved surfaces can only be treated by the method mentioned on p. 11. At the point of incidence a tangent plane is drawn, and for that particular incident ray the reflection is considered as taking place at that tangent plane. Fortunately, however, there are much less tedious methods of dealing with reflection at spherical surfaces, with which alone we are now concerned; these we proceed to describe in their simplest form.

Concave Spherical Mirrors; Axial or Centric Pencils.—Let AK represent a concave mirror, its surface being part of a sphere of which the centre is at C (Fig. 9). Any line drawn through C to the mirror is called an axis of the mirror, and when the parts of the mirror on either side of the axis are symmetrical it is called the Principal Axis. In the figure CA is the principal axis, and A the vertex of the mirror. Let P be a luminous point on the axis distant PA or p from the mirror, and consider the incident ray PK when the point K is near the vertex A. Join CK and make the angle CKQ numerically equal to the angle of incidence CKP. If now the figure be supposed to revolve round the principal axis PA, it is clear that PK will trace out the limits of a thin axial cone of incident light, while the point K will trace out a narrow circular zone on the mirror, and the reflected rays from this zone must all intersect the axis at the point Q. Consequently Q must be the image of P as formed by reflection at this narrow zone. Let CA, the radius of the mirror, be denoted by r, and let QA be denoted by q.

Now, since in the triangle PKQ the vertical angle is bisected by KC, which cuts the base at C—

$$\frac{PK}{QK} = \frac{PC}{CQ} = \frac{PA - CA}{CA - QA} = \frac{p - r}{r - q} \quad \text{(Euc. vi. 3)}$$

When K is very near the vertex A, PK and QK will be almost identical with PA and QA; under these circum-

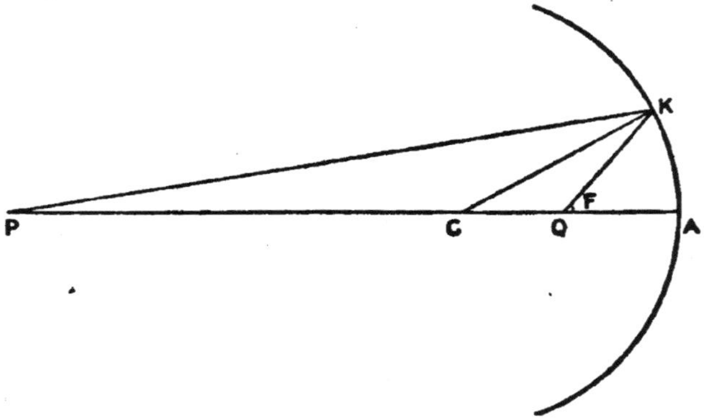

Fig. 9.

stances we may replace PK and QK by p and q, and so we get the following formula for a very thin centric pencil:—

$$\frac{p}{q} = \frac{p - r}{r - q}$$

$$\therefore pr - pq = pq - rq \quad \text{or} \quad qr + pr = 2pq$$

On dividing by pqr we obtain the formula

$$\frac{1}{p} + \frac{1}{q} = \frac{2}{r}$$

The point Q marks the situation of the real image of P; it is a real image because it is formed by the intersection of the reflected rays themselves (see p. 13). Again, Q is often called the *conjugate focus* of P. It is termed a focus (fireplace) because the place where heat rays intersect is hotter than any other. If, for instance, the mirror be used to form an image of the sun, a piece of paper placed at the focus will be burnt up. The term "conjugate" implies that if the object P be placed at Q, the image will then be in the old situation of P. In fact, P and Q are mutually convertible whenever the image formed at Q is real.

We will now consider the formula a little more closely. If PA or p be diminished until $p = r$, QA or q becomes greater until q becomes equal to r, for

$$\frac{1}{q} = \frac{2}{r} - \frac{1}{p} = \frac{2}{r} - \frac{1}{r} = \frac{1}{r}$$

This only means that when the luminous point is placed at C all the incident light reaches the mirror in the direction of its radii, or the normals to the surface, so that the reflected light must travel back by the same path; the image Q is then coincident with the object at C.

If the object P be brought still closer to the mirror (between C and F), the image Q will be formed at a greater distance from it; in fact, the situations of P and Q will be interchanged.

Now let us suppose the object P to be removed to a great distance; the image Q will be formed nearer the mirror. In the limiting case when the distance of P is infinite, let us see what our formula tells us. Here $p = \infty$.

$$\text{so } \frac{1}{\infty} + \frac{1}{q} = \frac{2}{r}. \quad \text{But } \frac{1}{\infty} = 0, \quad \text{so } \frac{1}{q} = \frac{2}{r}$$

This means that if the object is at an infinite distance, or, what comes to the same thing, if the incident beam consist of parallel rays, the corresponding focus is at a point F such that FA = $\frac{1}{2}$CA. This point F is called the **Principal Focus**, and the distance FA is usually symbolized by f. Clearly, as the direction of light is reversible, it follows that if a luminous point be placed at F the reflected rays will form a beam of light parallel to the axis.

Since we now know that $f = \frac{r}{2}$, we may give the previous formula in its more usual form—

$$\frac{1}{p} + \frac{1}{q} = \frac{1}{f} \quad \text{or} \quad \frac{f}{p} + \frac{f}{q} = 1$$

What will happen if the object be placed nearer the mirror than F? Our formula will tell us. Suppose that $f = 4$, and that $p = 3$;

then $$\frac{1}{q} = \frac{1}{f} - \frac{1}{p} = \frac{1}{4} - \frac{1}{3} = -\frac{1}{12} \qquad \therefore q = -12$$

QA or q is then -12. What is the meaning of the negative sign? We have regarded the direction from left to right (*e.g.* PA and CA) as positive; the opposite direction, from right to left, must therefore be the negative direction. Consequently, QA must now be measured in this direction, that is, the point Q must be behind the mirror. The reader is urged to draw for himself a diagram illustrating this case; that is, let him draw an arc of 8 cm. radius with its focus at 4 cm. from the mirror, and from a point P, 3 cm. from the mirror, let him draw any incident ray, PK, to the mirror; let him then mark a point Q, 12 cm. to the right of (behind) the mirror, and join QK and produce it. He will find that the normal CK bisects the angle between PK and the prolongation of QK, showing that KQ is the prolongation of the reflected ray in this case—in other words, Q is a virtual image of P, as it is formed, not by the reflected ray itself, but by its prolongation (see p. 13).

Whenever Q is a virtual image, P and Q are no longer interchangeable; in this case, for instance, the object if placed in the situation of Q would be behind the reflecting surface of the mirror, so no image would be formed.

Signs.—This case will indicate the importance to be attached to the meaning of algebraic signs. It will be found, if due attention be paid to them when thin axial pencils are being considered, that there are only two formulæ that need be remembered for reflection or refraction at single spherical surfaces, and for lenses of any kind or for any combination of them. The essential thing is to be *consistent* during any calculation; any inconsistency may lead to totally erroneous results.

In all optical problems it is most important to remember

that the symbol PC for a line, such as that in Figs. 9 and 10, denotes the distance from P to C, so that it not only expresses a length, but also the direction in which that length is measured. Euclid uses the term PC as identical with CP. We, however, must regard PC as a vectorial symbol, and therefore equal to $-$CP. Consequently—

$$PC = PA + AC = PA - CA = AC - AP$$

Similar vectorial expressions will be used throughout this book.

We have adopted the usual conventions that directions from left to right are considered positive, and those from right to left negative. Further, we shall regard directions from below upwards as positive, and those from above downwards as negative. As regards angles, we shall designate the directions of rotation in the usual way. All rotations in the counter-clockwise direction are considered positive, while all clockwise rotations are considered negative. The term "Standard Notation" will in this book be used to denote this device of signs to indicate these various directions.

In optics more blunders are due to the neglect of the meaning of signs than to any other cause, so it is well worth while devoting some attention to the subject. When a correct mathematical formula is given, one knows that it must be universally true, whatever values and whatever signs are given in a special case to the algebraic symbols in the formula.

Convex Spherical Mirrors.—When the mirror is convex, the point C, the centre, is behind the mirror, so that CA is negative. A mathematician would know without any further proof that the formula $\frac{1}{p}+\frac{1}{q}=\frac{2}{r}$ must be true if the appropriate sign were given to r. As, however, all our readers are not necessarily mathematicians, we will give a formal proof of this case.

As before, let C (Fig. 10) be the centre of curvature of the convex mirror AK, and let the incident ray PK be reflected at K in the direction KR, so that the angle of incidence $\phi = -\phi'$. Produce RK to Q, cutting the axis in Q. Then PKQ is a triangle, of which the exterior angle PKR is bisected by the line CK, that meets the base produced in C.

$$\therefore \frac{PK}{QK} = \frac{PC}{CQ} = \frac{PA + AC}{CA + AQ} = \frac{PA - CA}{CA - QA} = \frac{p - r}{r - q} \quad \text{(Euc. vi. A)}$$

When, therefore, a thin centric pencil is under considera-

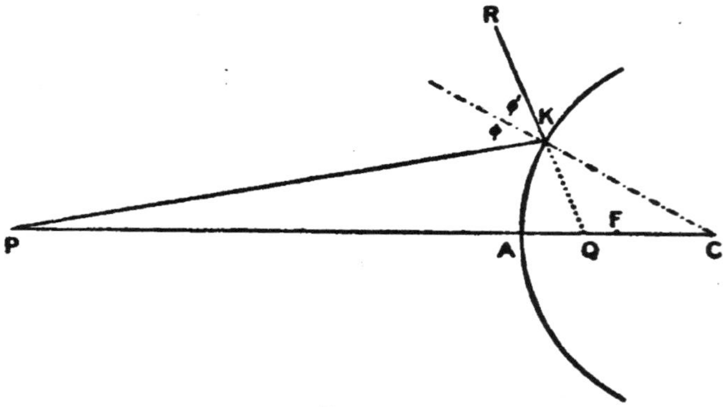

Fig. 10.

tion, and PK and QK may be regarded as equal to PA and QA, or p and q, we have, as before—

$$\frac{p}{q} = \frac{p - r}{r - q}, \text{ which reduces to } \frac{1}{p} + \frac{1}{q} = \frac{2}{r}$$

When the incident rays are parallel, *i.e.* when $p = \infty$, $\frac{1}{p} = 0$, consequently $\frac{1}{q} = \frac{2}{r} = \frac{1}{f}$, so they are reflected as if they came from a point F behind the mirror, such that FA $= \frac{1}{2}$CA; in other words, the principal focus is virtual and f is negative. We see, then, that the old formula $\frac{1}{p} + \frac{1}{q} = \frac{2}{r} = \frac{1}{f}$ still holds good if we assign the proper negative value to r and f.

Ex.—An object is placed 8 ins. in front of a spherical mirror, and an image of it is formed 4·8 ins. behind the mirror. What is the radius of curvature of the mirror, and what is its focal length?

In this case $p = 8$ ins. and $q = -4{\cdot}8$ ins., for the image is behind the mirror. So we get, on substituting these numerical values for the symbols, and paying due regard to the signs they bear—

$$\frac{1}{p}+\frac{1}{q}=\frac{2}{r}=\frac{1}{f}, \text{ or } \frac{1}{8}+\frac{1}{-4{\cdot}8}=\frac{2}{r}, \text{ or } \frac{1}{f}=\frac{6-10}{48}=\frac{1}{-12}$$

$$\therefore f = -12 \text{ and } r = -24$$

The negative signs show that both the radius and the focal distance are to be measured in the negative direction, *i.e.* both C and F are behind the mirror, that is the mirror is convex.

This example will show the wealth of information that is contained in this simple formula.

Geometrical Construction of the Image.—Fig. 11 represents a concave mirror, the centre of which is at C, and an object, AB, on the principal axis of the mirror; we are now going to show how the image *ab* can be constructed, assuming, of course, that only centric pencils contribute to its formation. As has already been pointed out, the previous formula is only true for thin centric or axial pencils, consequently in this case we shall draw a tangent plane HOH′ to the vertex O of the mirror, which will be a better representation of this centric portion than the whole curved surface of the mirror would be. In the figure, the line AC*a* may be taken to represent a thin centric pencil as it passes through the centre C, but it lies on a secondary, not on the principal, axis, and for this reason the term "centric" is less likely to be misunderstood than "axial." The plane HOH′ will in future be called the Principal Plane. We will give two methods by which the image *a* can be found of a point A that is not on the principal axis.

(A) *Point not on the Principal Axis.*—

(1) Draw AC*a* through the centre C, and draw AH to the principal plane parallel to the principal axis; draw HF*a* through the focus to meet the line AC*a* in *a*.

Then *a*, the point of intersection of HF*a* and AC*a*, marks the position of the image of the point A.

(2) (Dotted lines.) Draw AC*a* through C, and draw AFH′ through the focus to the principal plane; draw H′*a* parallel to the principal axis until it meets AC*a* in *a*. The point *a* is the image of the point A.

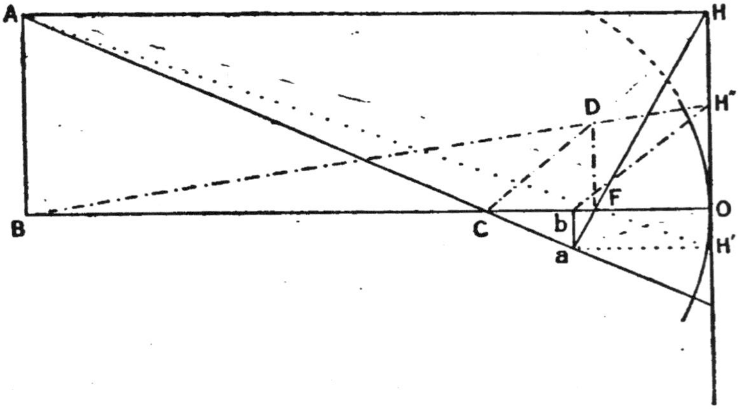

FIG. 11.

(1) The reason of this procedure will be apparent from the following considerations. We know that any ray parallel to the principal axis will be reflected through F, the principal focus; consequently, if AH represent such a ray it must be reflected as HF*a*. Further, we know that any ray drawn through the centre C must be reflected back along the same course through C. Hence the point of intersection *a* must represent the image of A. It is true that the line AH does not represent any ray that is actually incident on the mirror, for the point of its incidence would be so eccentric that it would not be reflected through F. For this reason the eccentric part of the mirror (suggested by spaced lines)

has been covered up. We are, however, justified in asserting that a small centric pencil from A will come to its conjugate focus at the point of intersection *a* of HF and AC.

(2) We also know that light from any luminous point at F must be reflected back parallel to the principal axis; consequently the ray FH′ must be reflected back as H′*a*, so that *a*, its point of intersection with AC, must be the image of A.

A vertical plane at F perpendicular to the principal axis, such as FD, is called the Focal Plane, and has the following properties: Light from any point on this plane will after reflection travel in rays parallel to that axis on which the point lies. For instance, light from D will be reflected from the mirror as a beam of parallel rays in the direction H″*b*, which is parallel to DC, the axis on which D lies. Also all pencils of parallel rays that are but slightly inclined to the principal axis will after reflection intersect in some point on the focal plane. This property enables us to determine the position of the image of a point that lies on the principal axis.

(B) *Point on the Principal Axis.* (Spaced and dotted lines.)

Through B draw any ray BDH″, cutting the focal plane in D; join DC and draw H″*b* parallel to DC, cutting the principal axis in *b*. Then the point *b* is the image of the point B.

Precisely the same construction can be applied to the case of a convex mirror, which the reader is recommended to draw for himself. It will be noted that whenever the object is vertical, the image also will be vertical, so that in drawing the image of AB in practice, all one has to do is to find the position of *a* by either of the methods given, and then draw a vertical line from it to meet the principal axis in *b*.

Size of the Image.—From a consideration of Fig. 11 it is easy to find an expression for the height of the image (*i*) as compared with the height of the object (*o*).

(1) Noting that ba is equal to OH′, we find by similar triangles that

$$\frac{i}{o}=\frac{ba}{\mathrm{BA}}=\frac{\mathrm{OH'}}{\mathrm{BA}}=\frac{\mathrm{FO}}{\mathrm{FB}}=\frac{\mathrm{FO}}{\mathrm{FO}-\mathrm{BO}}=\frac{f}{f-p}$$

(2) And, seeing that BA is equal to OH, it follows that

$$\frac{i}{o}=\frac{ba}{\mathrm{BA}}=\frac{ba}{\mathrm{OH}}=\frac{\mathrm{F}b}{\mathrm{FO}}=\frac{\mathrm{FO}-b\mathrm{O}}{\mathrm{FO}}=\frac{f-q}{f}$$

The formulæ which we have now proved, $\frac{f}{p}+\frac{f}{q}=1$, dealing with the position of the image, and these

$$\frac{i}{o}=\frac{f}{f-p}=\frac{f-q}{f},$$

dealing with the size of the image, can hardly be overrated, since almost exactly the same formulæ will be found to be true when we come to deal with refraction at spherical surfaces, while the method of construction of the image and the method of determining its size are merely a repetition of what we have just done.

It should be noted that *all erect images are virtual, while all inverted images are real.*

An example or two will show the value of these formulæ.

Ex. (1) A concave mirror has a radius of curvature of 10 ins. What is its focal length? An object $4\frac{1}{2}$ ins. in height is placed 50 ins. in front of the mirror. What is the height of the image, and where is it formed?

$$\text{Here } r=10, \text{ so } f=\frac{r}{2} \text{ or } 5 \text{ ins.}$$

$$\text{And since } \frac{f}{p}+\frac{f}{q}=1,\ \frac{f}{q}=1-\frac{f}{p}=\frac{p-f}{p}\quad\therefore q=\frac{fp}{p-f}$$

(This is the best form to use in all cases, as a similar form holds good for refraction at spherical surfaces.)

$$\text{Then}\qquad q=\frac{5\times 50}{50-5}=\frac{50}{9}=5\tfrac{5}{9}\text{ ins.}$$

The position of the image is then $5\frac{5}{9}$ ins. in front of the mirror, since q is positive.

Again, since $\dfrac{i}{o} = \dfrac{f}{f-p}$, $\dfrac{i}{o} = \dfrac{5}{5-50} = \dfrac{5}{-45} = -\dfrac{1}{9}$

The image is therefore inverted (as the sign is negative) and real, and it is $\frac{1}{9}$ the height of the object; *i.e.*

$$i = -\tfrac{1}{9}(4\tfrac{1}{2}) = -\tfrac{1}{2} \text{ in.}$$

It is immaterial which formula for the size of the image we use. Let us try the other formula:

$$\frac{i}{o} = \frac{f-q}{f} = \frac{5-\frac{50}{9}}{5} = \frac{-5}{5 \times 9} = -\frac{1}{9}$$

So the height of the image is $-\frac{1}{9}(4\frac{1}{2})$, or $-\frac{1}{2}$ in.

Of course, the formula about size refers also to width. Our second example is slightly more difficult.

Ex. (2) An object 6 cm. in width is placed at a distance of 9 cm. from a reflecting surface. A virtual image 2·4 mm. in width is formed of it. What is the radius of curvature of the reflecting surface?

$$\text{Since } \frac{i}{o} = \frac{f}{f-p}, \quad if - ip = fo \quad \therefore f(i-o) = ip$$

and since the image is virtual, i is positive.

Consequently

$$r \text{ or } 2f = \frac{2ip}{i-o} = \frac{2 \times 2{\cdot}4 \times 90}{2{\cdot}4 - 60} = \frac{4{\cdot}8 \times 90}{-57{\cdot}6}$$

$$r = -\frac{48}{6{\cdot}4} = -\frac{60}{8} = -7{\cdot}5 \text{ mm.}$$

As the sign is negative, the surface must be convex.

It may be noted that this is the basis of the method by which the radius of curvature of the cornea is determined in a living subject. A special apparatus is used to measure the size of the image reflected from the surface of the cornea, and from this measurement the curvature is calculated as in this example.

Graphic Method for Spherical Mirrors.—There is a very simple graphic method for finding the position and relative size of the image, which we will now give. The formula $\frac{f}{p} + \frac{f}{q} = 1$ is so similar in form to the well-known equation to a line in terms of its intercepts $\frac{x}{a} + \frac{y}{b} = 1$, that it at once suggests the following device (Fig. 12).

Draw two axes PH and HQ at right angles to each other, and consider H the origin. On PH mark off a distance F′H equal to the focal length f of the mirror, and draw the line F″F′ also equal to f at right angles to PH.

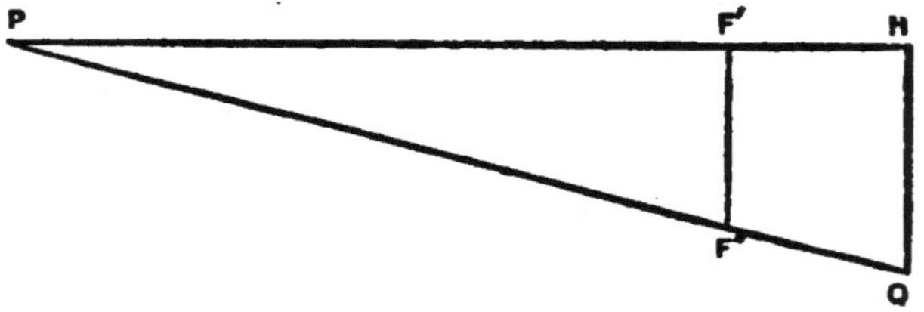

FIG. 12.

As with a concave mirror f is positive, F′H and F″F′ must be measured in the positive directions, *i.e.* either from left to right or from below upwards (standard notation).

Suppose that an object is placed 20 ins. from a concave mirror of focus 4 ins., and we wish to determine the position, size, and nature of the image. Mark off the point P on PH so that PH = 20; join PF″, cutting HQ in Q.

Then QH is the distance of the image from the mirror; as it is measured from below upwards QH is positive, and it is found to be 5 ins. in length, so the image is situated 5 ins. in front of the mirror.

Again, $\frac{i}{o} = \frac{f}{f-p}$; but $f - p =$ F′H − PH = F′P

$$\therefore \frac{i}{o} = \frac{\text{F}'\text{H}}{\text{F}'\text{P}}, \text{ which in this case is } \frac{4}{-16} = -\frac{1}{4}$$

As F′H and F′P are measured in opposite directions $\frac{i}{o}$

must be negative, so the image is inverted and real, and its height is $-\frac{1}{4}$ of that of the object.

With the convex mirror f is negative, let $f = -4$ ins., so F′H and F″F′ are drawn as shown in Fig. 13, in the negative direction. Suppose the object to be 16 ins. from the mirror, PH is then made 16, and QH is found to be $-3 \cdot 2$ (negative because measured downwards). Consequently the image is situated 3·2 ins. behind the mirror.

And as $$\frac{i}{o} = \frac{F'H}{F'P} = \frac{-4}{-20} = \frac{1}{5}$$

the image is virtual and erect, because $\frac{i}{o}$ is positive (F′H and

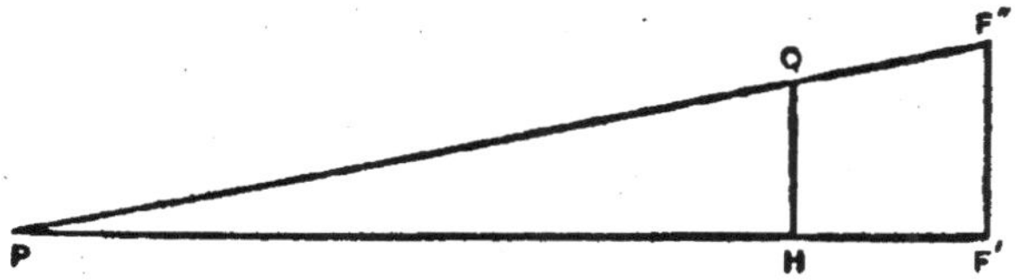

FIG. 13.

F′P being measured in the same direction), and its height is $\frac{1}{5}$ that of the object.

***Eccentric Pencils; Spherical Aberration.**—In the preceding sections we have been considering the reflection of small centric pencils only; when the incident light forms a wide cone the several reflected rays cross each other at different points, as is indicated in the diagram (Fig. 14). Paying attention first to the eccentric pencil that falls on the mirror at PQ, the reflected rays cross at F_1 and meet the axis in a line at F_2 nearer the mirror than I, the conjugate focus of S for centric pencils. The figure, of course, only represents a section of the whole mirror; if we consider it to be rotated through a small angle about the axis OCS, the point F_1 will trace out a small arc (approximately a line) so that at F_1 and F_2 two small lines will be formed: these are called the primary and the secondary focal lines.

It will be noticed, also, that all the reflected rays that are incident on the mirror from S touch a certain caustic surface which has a cusp at I. This caustic curve may be commonly observed on the surface of the fluid in a teacup, being formed by the light that is reflected from the inside of the cup. A caustic curve is frequently defined as the locus of the primary focal lines; this only means that it is formed by the points of intersection of successive reflected rays.

The term "*spherical aberration*" is applied to all the phenomena depicted in Fig. 14. If the reflecting surface had not been spherical, but had been part of an ellipsoid of revolution of which S and I were foci, all the light from S that fell on this elliptical surface would have been accurately

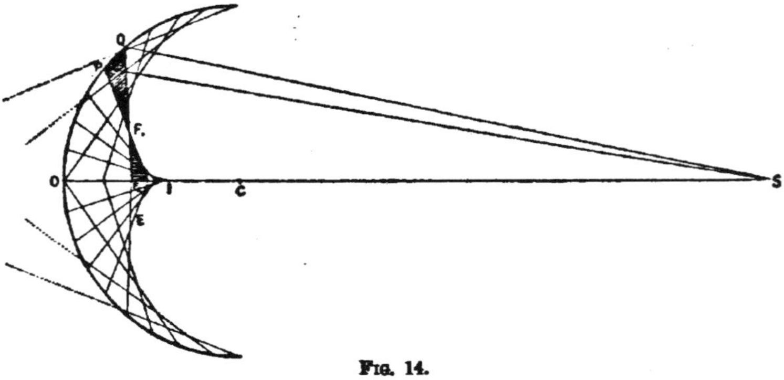

FIG. 14.

reflected to the conjugate focus I, and the surface would then be an "*aplanatic*" (ἀπλανής, not wandering) surface for the source S. Similarly, a paraboloid would be aplanatic for parallel rays.

The "*longitudinal aberration*" (IF_2) of a thin eccentric pencil is the distance between the point where it cuts the axis and the focus for a thin centric pencil.

Fig. 15 gives another view of the reflected pencil which is shown in section as the shaded area in Fig. 14. The incident cone is supposed to be pyramidal, so that the reflecting portion of the mirror concerned is of the rectangular shape $PQRR_1$. The reflected pencil is of a very curious shape, the upper and lower rays (*e.g.* QF_1 and PF_1) of the two sides intersect in the primary focal line at F_1, while the rays QF_2 and RF_2 from the upper edge intersect at F_2 the bottom of the secondary focal line, and PF_2 and R_1F_2 from the lower edge at the top of the secondary focal line. Pencils like this which do not converge to a point but to two focal lines are called "astigmatic pencils" (α priv., στίγμα, a point). As

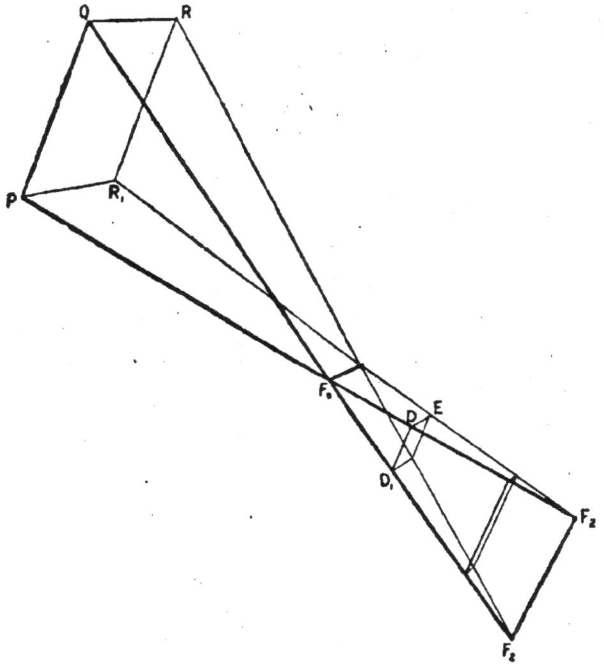

FIG. 15.

we shall find that precisely the same condition arises with eccentric pencils when refracted, it is well to spend some little time in forming a clear conception of the solid figure represented between the two focal lines at F_1 and F_2. In crystallography this sort of double wedge is called a sphenoid. Near F_2 its section would be oblong, its height much greater than its width; at D the section would be square, while nearer F_1 it would again be oblong, but with its width greater than its height. Clearly if the incident pencil were circular in section, the outline at D would be a circle. D is then the position of what is called "the circle of least confusion," and it is the best representative of the focus of an astigmatic pencil.

We will conclude this chapter by giving the formulæ without proof by which the position of the primary and secondary focal lines can be determined of an incident thin oblique or eccentric pencil on a spherical surface (Appendix, p. 109). Let SP be denoted by u, F_1P by v_1, and F_2P by v_2, and the angle of incidence by ϕ.

Then $\frac{1}{u}+\frac{1}{v_1}=\frac{2}{r\cos\phi}$ and $\frac{1}{u}+\frac{1}{v_2}=\frac{2\cos\phi}{r}$

QUESTIONS.

(1) An object 9 cm. in height is placed 10 cm. in front of a concave mirror of focal length 25 cm. What is the height and character of the image, and where is it formed?

(2) The radius of a concave mirror is 16 ins. What is the distance of the image from the mirror when the object is placed at a distance of 12 ins., and when placed at a distance of 4 ins.?

(3) A virtual image one-fourth the height of the object is formed by a mirror. If the distance of the object be 9 ins. what is the radius of the mirror?

(4) An inverted image of a candle is thrown on a screen at a distance of 6 feet from a mirror of focal length 6 ins. Where is the candle placed, and what is the relative size of the image?

CHAPTER IV

REFRACTION AT PLANE SURFACES

So far we have been considering the path taken by light as it travels through a single homogeneous medium; we must now find out what happens when light passes through more than one medium. For the sake of simplicity, we shall limit ourselves to the consideration of homogeneous light, *i.e.* light of the same wave frequency or colour, traversing different homogeneous media. When light passes from one medium into another, *e.g.* from air into glass, its course is altered, and the light is said to be refracted. The subject was experimentally investigated in 1621 by Willebroard Snell, who found that there are two laws governing refraction:

LAW 1. The refracted ray lies in the plane of incidence.

LAW 2. The sines of the angles of incidence and refraction are in a constant ratio for the same two media.

The first law requires no explanation after what we have already said about the similar law of reflection (p. 11).

Students of physics will know that light *in vacuo* travels at a rate of more than 186,000 miles per second, and that in dense media its velocity is less. Indeed, in Optics the expression dense medium merely means a medium in which light travels with a relatively slow velocity. For instance, water is dense as compared with air, but rare as compared with glass, the speed of light in water is about $\frac{3}{4}$ of its speed in air, but about $\frac{9}{8}$ of its speed in glass. Now, it has been mathematically proved and experimentally demonstrated that the constant ratio in Snell's law is identical with this velocity ratio. If we denote the angle of incidence by ϕ and that of refraction by ϕ', when light passes from water into air its course is so altered that $\frac{\sin \phi}{\sin \phi'}, = \frac{V_w}{V_a}$, *i.e.* about $\frac{3}{4}$. When light passes from air into glass, $\frac{\sin \phi}{\sin \phi'}, = \frac{3}{2}$ approximately, for the speed of light in air is about $\frac{3}{2}$ times its speed in glass.

Let BB′ (Fig. 16) represent the vertical plane that forms the boundary of a dense medium, glass for instance, and let SC be an incident ray, CQ the corresponding refracted ray, and let NCM denote the normal at the point of incidence C. The original direction of the incident light is indicated by the dotted line CS′, so MCS′ is the angle of incidence ϕ, and MCQ is the angle of refraction ϕ'. Note that the angles of incidence and refraction are to be measured from the same normal, in this case CM. In the figure $\frac{\sin \phi}{\sin \phi'}, = \frac{S'L}{QM}$, and they are both positive, as of course they should be, for the ratio $\frac{\sin \phi}{\sin \phi'}$, is the ratio of the speed of light in the first medium to its speed in the second medium. As light is reversible, we see that an incident ray in the direction QC in the dense medium will be refracted at the surface BB′ of the rare medium as CS. Indeed, it follows from Law (2) that *all light on entering a dense medium is refracted towards the normal, and on entering a rare medium away from the normal.*

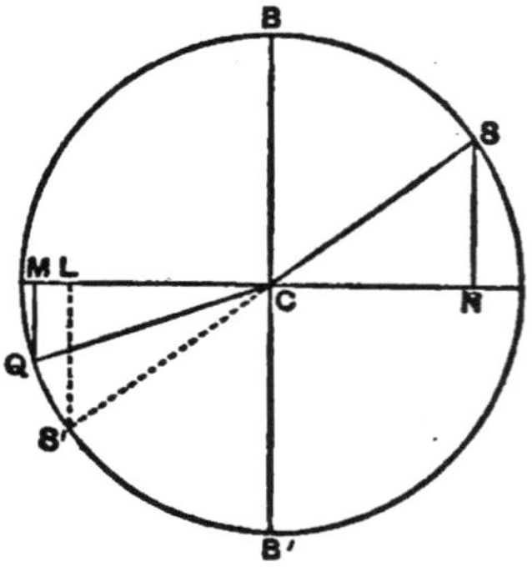

Fig. 16.

The ratio of the velocity of light in a vacuum to its velocity in a certain medium is termed the absolute index of refraction of that medium, and it is denoted by the symbol μ. Thus, if the absolute index of water be $\frac{4}{3}$ and that of a certain kind of glass be $\frac{3}{2}$, it is easy to determine the relative index of water and glass, or $\frac{V_w}{V_g}$.

$$\frac{V_w}{V_g} = \frac{V_0}{V_g} \times \frac{V_w}{V_0} = \frac{3}{2} \times \frac{3}{4} = \frac{9}{8} \text{ or } {}_w\mu_g = \frac{\mu_g}{\mu_w}$$

Snell's law may be then stated as $\frac{\sin\phi}{\sin\phi'} = \frac{\mu_2}{\mu_1}$, where the subscripts denote the first and second media; when the first medium is air, $\frac{\sin\phi}{\sin\phi'} = \mu_2$. The refractive index of air differs so slightly from unity (being less than 1·0003) that we shall commonly regard it as unity and denote it by μ_0.

Returning again to Fig. 16, the angle S′CQ is called the **deviation** (D) of the light in this medium when the angle of incidence or ϕ is MCS′. As SCS′ is the original course of the light, the deviation S′CQ or D is equal to − (MCS′ − MCQ) or $D = -(\phi - \phi')$. This is clear, for the new direction ϕ' must be equal to (the old direction) $\phi + D$. Annexed is a table in which the angles of refraction and deviation are given that correspond with certain angles of incidence at the surface of a kind of glass whose index of refraction μ is 1·52, and again at the surface of water when μ = 1·333.

$\mu = 1·52$.			$\mu = 1·333$.		
ϕ	ϕ'	D	ϕ	ϕ'	D
20°	13° 0′	− 7° 0′	20°	14° 52′	− 5° 8′
40°	25° 1′	− 14° 59′	40°	28° 50′	− 11° 10′
60°	34° 44′	− 25° 16′	60°	40° 31′	− 19° 29′
80°	40° 23′	− 39° 37′	80°	47° 38′	− 32° 22′
90°	41° 8′	− 48° 52′	90°	48° 36′	− 41° 24′

It will be noticed that when the angle of incidence increases uniformly, as in the first four rows, the angle of refraction increases slower and slower; consequently, the deviation ($\phi' - \phi$) increases faster and faster. It follows from what has been said before that if the light be considered as passing from the dense medium to the rare medium, we must consider the ϕ' in the above table as the angle of incidence, and ϕ the angle of refraction.

What will happen when the angle of incidence in water, for instance, is greater than 48° 36′? When ϕ is nearly 48° 36′ (Fig. 17) the refracted light just skims along the surface of the water. If ϕ be more than 48° 36′ the light will be unable to leave the water, and it will be *totally reflected.* The angle of incidence (48° 36′) at which this phenomenon of total reflection occurs is called the critical angle. It should be noted that when the path of light is from a

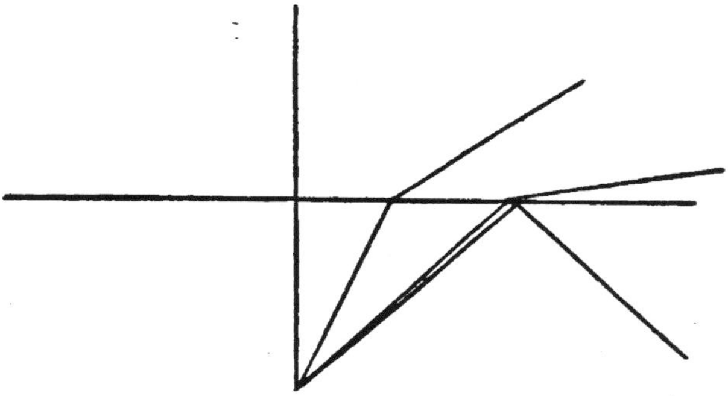

FIG. 17.

dense to a rare medium the relative index of refraction is less than unity, for ${}_d\mu_r = \frac{\mu_r}{\mu_d}$; in the case where $\mu_r = 1$, ${}_d\mu_r = \frac{1}{\mu_d}$. The critical angle is easily found for any medium; we have merely to give ϕ' its maximum value of 90°, and our formula tells us that $\frac{\sin \phi}{\sin 90°} = {}_d\mu_r = \frac{1}{\mu_d}$, $\therefore$ the critical angle is $\sin^{-1}\left(\frac{1}{\mu_d}\right)$; thus $\sin^{-1}\left(\frac{1}{1{\cdot}333}\right) = 48° \, 36'$.

Total reflection is frequently taken advantage of in the construction of optical instruments, *e.g.* the camera lucida (Appendix, p. [illegible]) and prism binoculars. Similarly when the surface of water is viewed in a glass held above the level of the eye, the silvery brilliance of the surface is due to total reflection. In aquariums the surface of the water is often seen to act as a brilliant mirror.

We see, then, that part at any rate of the light in a rare medium can always enter an adjoining dense medium, but that when light in a dense medium is incident at any angle greater than the critical angle, none of it will leave the dense medium, as it will be totally reflected at the boundary of the rare medium.

Image by Refraction at a Plane Surface.—Let P (Fig. 18) be a source of light in a dense medium (μ') that is separated

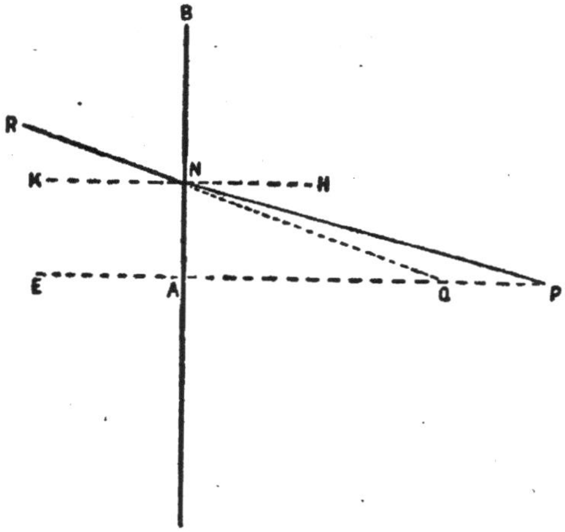

FIG. 18.

by a plane surface AB from a rare medium (μ_0), say air. Draw PAE normal to the surface. We will consider how the point P will appear to an eye at E.

Since P is sending out light in all directions, we may regard PA and PN as two of them. The ray PA will undergo no alteration, for since PA is normal, both ϕ and ϕ' must be 0. Draw the normal HNK at the point of incidence N; then HNP = ϕ, and if HNQ be so drawn that $\frac{\sin \text{HNP}}{\sin \text{HNQ}} = \frac{\mu_0}{\mu'}$, HNQ must be ϕ'. Under these circumstances produce QN to R, then the ray PN will be refracted as NR. Now, if the figure be rotated round EP as axis, PN will trace out the limits of the axial incident cone and QNR the limits of the emergent cone. Since HNP and HNQ are equal to the alternate angles QPN and AQN,

$$\frac{\mu_0}{\mu'} \text{ or } \frac{1}{\mu'} = \frac{\sin \text{HNP}}{\sin \text{HNQ}} = \frac{\sin \text{QPN}}{\sin \text{AQN}} = \frac{\sin \text{QPN}}{\sin \text{NQP}} = \frac{\text{QN}}{\text{PN}}$$

As, however, we are only considering the narrow pencil that enters the pupil of the eye at E, the point N will be very close to A, and QN and PN may be regarded as equal to QA and PA; consequently, to an eye at E the position of the virtual image of P will be at Q when $\text{QA} = \frac{\mu_0}{\mu'}\text{PA} = \frac{\text{PA}}{\mu'}$.

If, for example, a small object in water is viewed from a point immediately above it, its apparent depth will be $\frac{3}{4}$ its real depth.

As in the case of reflection at spherical mirrors, this determination of the situation of the virtual image is only true when small pencils that are nearly normal to the surface are considered. The formulæ for oblique pencils are more complicated, as two focal lines are formed with a circle of least confusion between them (see Appendix, p. 111).

Refraction through a Plate.—Let us consider how an object P in air (μ_0) will appear to an eye E in air when viewed through a plate of glass (μ') of which the thickness t is AB (Fig. 19). The incident ray PN will be refracted as NM on entering the glass, and on again entering the air at M its course will assume the direction MR. As we are only considering what will be the position of the virtual image Q to an eye at E, we can find it very easily by the method of the last section.

At the surface AN the point P will form a virtual image at Q′ such that $\text{Q}'\text{A} = \frac{\mu'}{\mu_0}\text{PA}$. But now Q′ may be regarded as the object for the second refraction at BM, and $\text{Q}'\text{B} = \text{Q}'\text{A} + t$, consequently we have for the distance QB from the distal surface

$$\text{QB} = \frac{\mu_0}{\mu'}\text{Q}'\text{B} = \frac{\mu_0}{\mu'}(\text{Q}'\text{A} + t) = \text{PA} + \frac{\mu_0}{\mu'}t.$$

and since QA = QB $\mp t$,

$$QA = PA - t\left(1 - \frac{\mu_0}{\mu'}\right)$$

When $\mu_0 = 1$ and μ' is denoted by μ, $QA = PA - t\left(\frac{\mu - 1}{\mu}\right)$, and $QB = PA + \frac{t}{\mu}$. Thus if $\mu = \frac{3}{2}$, $QA = PA - \frac{t}{3}$, or the image is formed nearer the plate by $\frac{1}{3}$ of its thickness.

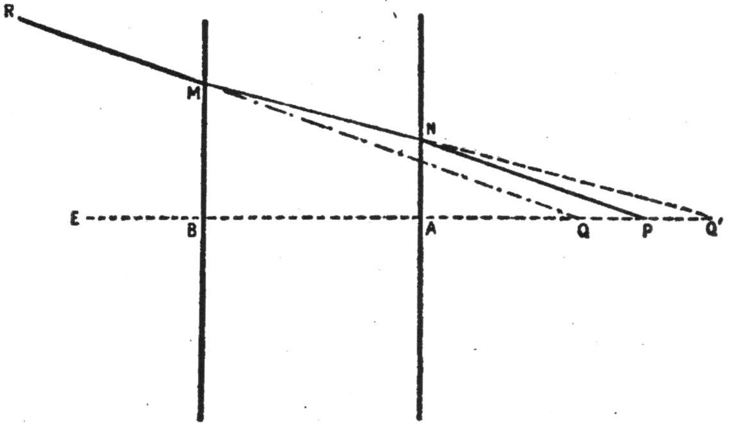

Fig. 19.

It will be noticed that the emergent ray MR is parallel to the incident ray PN; this is invariably true however many media there may be, provided that they are bounded by parallel planes, and provided that the final medium has the same refractive index as the initial medium.

Succession of Plates.—It is easy to see that, if any number of media bounded by parallel planes are in succession, light on emerging will pursue a course parallel to its original path if the initial and final media have the same refractive index μ_1. Let $\mu_1, \mu_2, \mu_3 \ldots$ denote the indices of the first, second, third . . . medium respectively, and let $\phi_1, \phi_2, \phi_3 \ldots$ denote the angles of incidence in these media; then

$$\frac{\sin \phi_1}{\sin \phi_2} = \frac{\mu_2}{\mu_1}, \ \frac{\sin \phi_2}{\sin \phi_3} = \frac{\mu_3}{\mu_2}, \text{ etc.}$$

for, since the media are bounded by parallel planes, the angle of refraction at one surface is equal to the angle of incidence at the next. Thus, if there are four media interposed (Fig. 20), and the angle of emergence into the final medium be denoted by θ,

$$\sin\phi_1 = \frac{\mu_2}{\mu_1}\cdot\frac{\mu_3}{\mu_2}\cdot\frac{\mu_4}{\mu_3}\cdot\frac{\mu_5}{\mu_4}\cdot\frac{\mu_1}{\mu_5}\sin\theta$$

$$\therefore \sin\theta = \sin\phi_1,$$

or, in other words, the final and initial rays are parallel. In

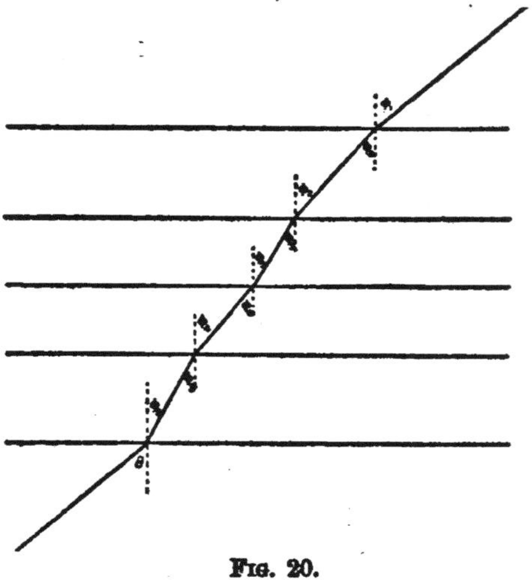

FIG. 20.

the diagram the third medium is represented as less dense than either of those adjoining it. Consequently, ϕ_4 is greater than ϕ_3 and ϕ_5.

Prisms.—Any refracting medium bounded by two plane surfaces which are inclined at an angle to each other is called a prism. The inclination of the faces BA and CA (Fig. 21) is called the refracting angle or the apical angle of the prism, and is usually denoted by A. The median vertical plane that bisects the apical angle is called the principal plane, while the plane CB at right angles to the principal plane is called the base of the prism. We proceed to demonstrate certain properties which are common to all prisms.

1. *When light passes through a prism which is denser than the surrounding medium, it is always deviated towards the base.*

If the face BM in Fig. 19 were rotated clockwise through a small angle about B, the plate would form a prism with its edge upwards and its base downwards; the angles of incidence and of refraction at BM will therefore diminish, and consequently any incident ray NM will on emergence be deviated away from MR towards the base. If the rotation of BM be continued, the angle of incidence at M, after passing through the value of 0, will change its sign and become positive, when a still more marked deviation of the emergent ray will occur. Were the face BM rotated counter-clockwise, the base of the prism then formed would be upwards, and at the same time the negative angle of incidence would increase, and hence also the angle of refraction. The deviation of the emergent ray would in such a case be upwards, towards the base.

If the prism be less dense than the surrounding medium the deviation is towards the edge of the prism.

II. *As the angle of a prism increases the deviation also increases.*

This follows immediately from the proof given of I.

III. *The apical angle is equal to the difference between the angles of refraction and incidence within the prism.*

When the apical angle A is measured in the same direction as the angle of deviation D,

$$A = \psi' - \phi'.$$

In all the books on Optics A is said to be equal to $\phi' + \psi'$, and in order to get this result a new and special convention is adopted for the signs attributed to ϕ' and ψ', which is never elsewhere employed. Our convention about clockwise and counter-clockwise rotation when consistently used, will be found to give results that always hold good.

Let SIRT denote a ray of light passing through the prism BAC (Fig. 21). Draw the normals at I and R, and let the angles of incidence and refraction at I be ϕ and ϕ', and the angles of incidence and emergence at R be ψ' and ψ. Note that in the figure ϕ and ϕ' are measured clockwise and are therefore negative, while ψ' and ψ are measured counter-clockwise and are consequently positive.

In the triangle AIR the angle IRA $= 90 - \psi'$, and the angle AIR $= 90 + \phi'$, for ϕ' is measured clockwise. Now,

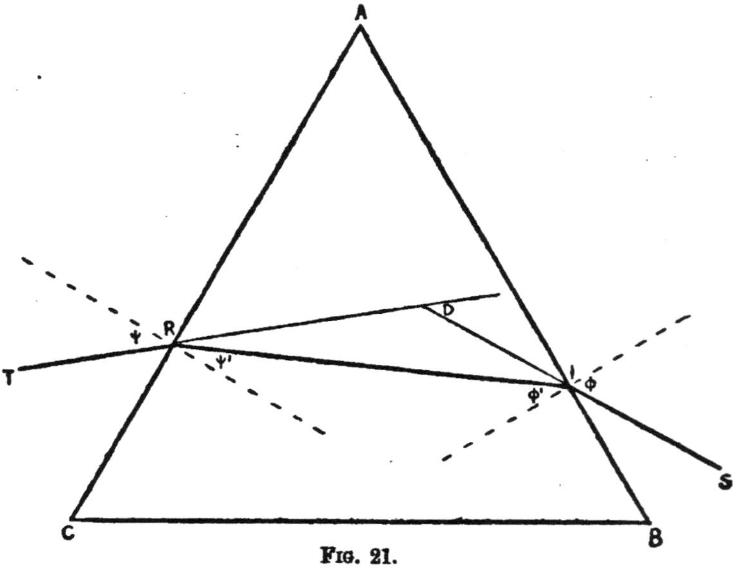

FIG. 21.

since the sum of the interior angles of the triangle AIR is 180°,

$$\text{RAI} + 90 - \psi' + 90 + \phi' = 180$$

$$\therefore \text{A or RAI} = \psi' - \phi'.$$

This is universally true whatever signs ϕ' and ψ' carry, whether they be different or both the same.

IV. *The total deviation is equal to the difference between the angles of emergence and incidence less the apical angle of the prism.* $D = \psi - \phi - A$.

If the rays SI and TR be produced to intersect at D, the exterior angle D of the triangle DRI is equal to the two interior and opposite angles IRD and DIR.

But $\text{IRD} = \psi - \psi'$, and DIR or $-\text{RID} = -(\phi - \phi')$

$$\therefore \text{D} = \psi - \phi - (\psi' - \phi') = \psi - \phi - \text{A}.$$

It is clear that the total deviation D is the deviation of the first surface, $-(\phi - \phi')$, added to the deviation of the second surface, $\psi - \psi'$.

As SIRT is the path of the light, the angle D is positive (counter-clockwise), and A is measured in the positive direction also (RAI). If the light were travelling in the direction TRIS the deviation D would be negative, and then A (or IAR) would be measured in the clockwise direction. As ψ and ϕ would then be interchanged, it would be found in that case also that

$$\text{D} = \psi - \phi - \text{A}.$$

V. *When a ray passes symmetrically through a prism, the deviation is a minimum.*

A ray passes symmetrically through a prism when the angle of incidence is numerically equal to the angle of emergence irrespective of sign, *i.e.* when $\phi = -\psi$.

If ϕ increases, ϕ' increases also; at the same time, however, ψ' diminishes, and consequently ψ also. But as the deviation $-(\phi - \phi')$ increases faster than the deviation $\psi - \psi'$ diminishes (p. 36), the total deviation must increase. If we consider the path of light reversed, it appears that when the angle of incidence is diminished the total deviation increases. Hence the symmetrical position is the position of minimum deviation.

When the prism is in the position of minimum deviation, $\psi = -\phi$ and $\psi' = -\phi'$;

$$\therefore \text{D} = 2\psi - \text{A}, \quad \text{and} \quad \text{A} = 2\psi'$$

$$\therefore \mu \text{ or } \frac{\sin\phi}{\sin\phi'} = \frac{\sin\psi}{\sin\psi'} = \frac{\sin\frac{1}{2}(\text{A}+\text{D})}{\sin\frac{1}{2}\text{A}}$$

This is the method used for determining the relative refractive index of any transparent substance. The angles D and A can be measured with great accuracy by means of a spectrometer, and from these data the value of μ can be found by this formula.

The prisms that are prescribed for spectacles are usually very weak, *i.e.* the angle A is rarely more than 5°, and for these a simple approximate formula can be given for the amount of deviation they induce. When ψ and ψ' are both very small, sin ψ and sin ψ' may be replaced by ψ and ψ', and the formula for the position of minimum deviation $D = 2\psi - A$ may be replaced by $2\mu\psi' - A$ or $(\mu - 1)A$, since $A = 2\psi'$. This is a most useful formula if only used for weak prisms when placed in the "symmetrical" position; thus, as the value of μ for glass is little more than 1·5, we find that the deviation produced by such a weak prism is about half its apical angle.

Image by Refraction through a Prism.—We have so far been only considering the course of light refracted by a prism; when we try to form some idea of the appearance of an object when viewed through a prism, we meet with several difficulties that will only be briefly mentioned here.

We have seen that the deviation is not proportional to the angle of incidence, and as an object of appreciable size presents many points from each of which light is falling on the prism at a different angle, it will be seen that the matter is a very complicated one.

In Fig. 22 an object P*p* is supposed to be viewed by an eye in the neighbourhood of RT, the narrow pencil from P that enters the eye being represented in the position of minimum deviation. Clearly a virtual image of P will be formed at Q, where the prolongations of the refracted rays intersect. It is obvious that the image will be displaced towards the edge of the prism, and that it will seem nearer by one-third the thickness of the prism traversed (p. 40). Now, if we consider light emerging from *p*, it is clear that if it fell on the prism in the position of minimum deviation it would form a virtual image in the neighbourhood of *q*. But an eye at R could not receive this pencil; the only light that would enter it must have had some such initial direction as *px*, and the final image of *p* would be indistinct, for the lowest ray of the pencil entering the eye from p will have undergone a smaller deviation than the uppermost one, and consequently a confused image of p will be formed about q'.

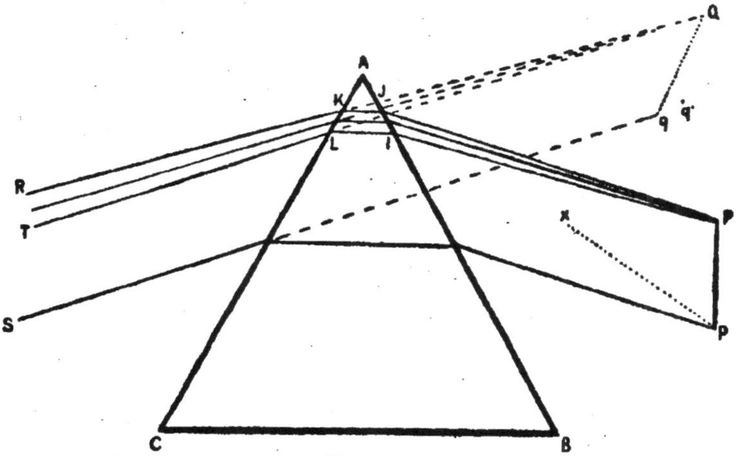

Fig. 22.

It will be convenient to give a summary of the appearances of a square object when viewed through a prism in different positions.

(1) When the plane of the prism is parallel to the plane of the object, the edge of the prism being upwards.

The image is raised above the level of the object, the sides being more raised than the mid-part, so that the upper and lower edges appear concave upwards.

(2) When the prism is rotated about a horizontal axis parallel to its edge.

The image rises, and the height of the image is either diminished or increased according as the edge of the prism is turned towards or away from the observer.

(3) When the prism is rotated about a vertical axis, its right side being turned away from the observer.

The right margin of the image is raised above its left margin, so that the right superior and the left inferior angles are rendered more acute. If the rotation be in the opposite direction, opposite results occur.

(4) When the prism is rotated about the sagittal line, *i.e.* about the visual line of the observer.

The image rotates in the same direction, for, being always displaced towards the edge, it follows it in its rotation. Let a point of light be supposed to fall on the centre of the circular screen depicted in Fig. 23. When a prism that causes a deviation R is interposed with its edge to the right, the point of light is deflected to the periphery of the circle on the right. If the prism be rotated through an angle ρ, the spot of light will move along the arc through the angle ρ. Consequently, the vertical displacement V will be R sin ρ, and the horizontal displacement H will be R cos ρ. A prism is frequently ordered by oculists in this way to correct a vertical and a horizontal deviation of the eye simultaneously. Clearly, if V and H are known the required prism can be found, for $R = \sqrt{V^2 + H^2}$. Practically, it is found what horizontal and vertical angular deviations are required by the patient. Say that they are θ and ϕ, then $\tan\theta = H$ and $\tan\phi = V$, and a prism of deviation D is ordered such that $\tan D = R$, and set at such an angle ρ that

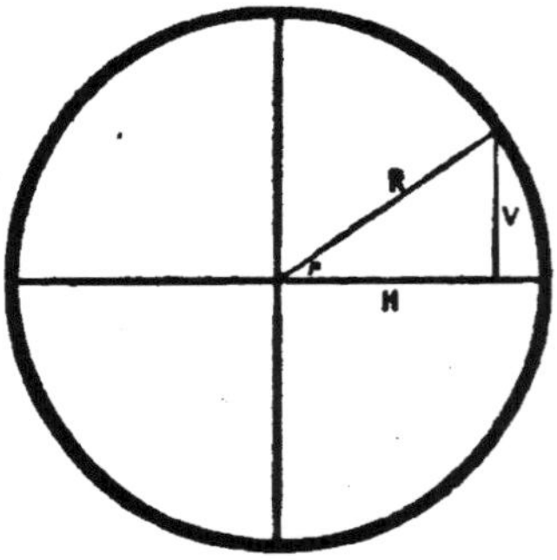

FIG. 23.

$$\tan\rho = \frac{V}{H} = \frac{\tan\phi}{\tan\theta}.$$

In ophthalmic practice the prisms are so weak that we may replace the tangents of θ, ϕ, and D by the angles (of minimum deviation) themselves.

The reader is urged to verify the statements made in this section by experiments with a prism, but a full explanation of his observations can only be obtained in more advanced treatises.

QUESTIONS.

(1) Show why a stick that is partly in and partly out of water appears bent at the surface of the water when viewed obliquely.

(2) It is found that when a plate of glass 7·7 mm. thick is placed over a microscopic object the microscope must be raised 2·7 mm. to bring the object into focus again. What is the refractive index of the glass?

(3) If the apical angle of a prism be 60°, and the minimum deviation for a certain kind of light 30°, what is the refractive index of the material of the prism for this light?

(4) A prism of small apical angle (2°), with refractive index 1·5, is placed in water of refractive index $\frac{4}{3}$. Show that its deviation is only about one-fourth of what it is in air.

(5) When viewing a distant object, each eye of a patient is found to deviate outwards 1° 44′ (nearly $\sqrt{3}$°), while the right eye deviates above the level of the fixation line of the left eye 1°. What two prisms would entirely relieve this defect?

CHAPTER V

REFRACTION AT A SPHERICAL SURFACE

WHEN dealing with reflection at a spherical surface (Fig. 9, p. 19), we considered the object P to be on the positive side of the mirror. It would clearly make no difference to the distance of the image from the mirror if the object were to the right instead of the left of the mirror, except that the image Q would then be situated to the right also. Since optical formulæ are universally true whatever values are given to p, q, and r, when *using* the formulæ it will generally be found convenient to regard the direction of the incident light as the positive direction. Hence PA or p may be regarded as positive whether measured from left to right or from right to left, and QA or q will be positive when both Q and P are on the same side of the surface of the medium, but q will be negative when Q and P are on opposite sides of the surface.

When *finding* any general formula, to avoid error, it is the simplest plan to use the "Standard Notation." As will be seen in the following two sections, there is no difficulty in obtaining a correct general formula when the object lies to the right of the refracting medium.

We shall in Fig. 24 (and sometimes in future) consider the object P to be situated to the right of the spherical arc AK to familiarize the reader with the fact that the direction in which light is travelling is quite immaterial when dealing with such questions as these.

Concave Surface of Dense Medium.—Let P be an object in

a medium whose refractive index is μ_0 distant PA or p from a concave spherical arc, AK, bounding a medium of index μ', and let CA or CK be the radius (r) of the arc; then, if CKP

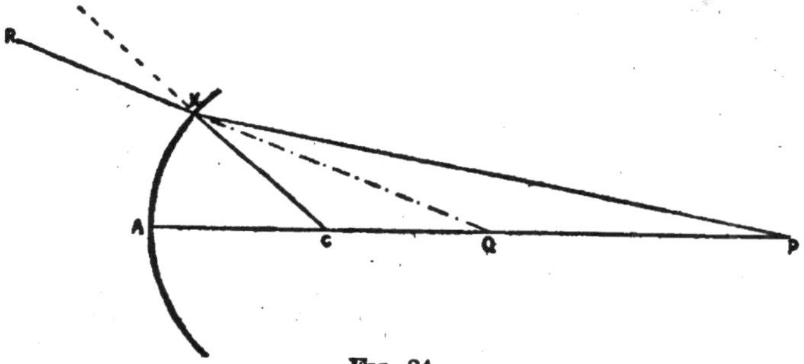

FIG. 24.

be the angle of incidence ϕ, and CKQ be the angle of refraction ϕ',

$$\frac{\mu'}{\mu_0} = \frac{\sin\phi}{\sin\phi'} = \frac{\sin \text{CKP}}{\sin \text{CKQ}}$$

Now, in the triangle PKC

$$\frac{\text{PC}}{\text{CK}} = \frac{\sin \text{CKP}}{\sin \text{KPC}} = \frac{\sin\phi}{\sin \text{KPC}}$$

and in the triangle CKQ

$$\frac{\text{QC}}{\text{CK}} = \frac{\sin \text{CKQ}}{\sin \text{KQC}} = \frac{\sin\phi'}{\sin \text{KQC}}$$

[Note that as PC, CK, and QC are all measured in the same direction, the angles CKP, CKQ, and KQC must all be measured in the same direction, whether positive or negative. In this case PC and CK are negative, so their ratio is equal to the ratio of the sines of two positive angles.]

On dividing the first expression by the second, we get

$$\frac{\text{PC}}{\text{QC}} = \frac{\sin\phi}{\sin\phi'}\cdot\frac{\sin \text{KQC}}{\sin \text{KPC}} = \frac{\mu'}{\mu_0}\cdot\frac{\sin \text{PQK}}{\sin \text{KPQ}} = \frac{\mu'}{\mu_0}\cdot\frac{\text{PK}}{\text{QK}}$$

i.e. $$\frac{\text{PA}-\text{CA}}{\text{QA}-\text{CA}} \quad\text{or}\quad \frac{p-r}{q-r} = \frac{\mu'}{\mu_0}\cdot\frac{\text{PK}}{\text{QK}}$$

This is universally true when $p = \text{PA}$, $q = \text{QA}$, and $r = \text{CA}$, and when μ_0 is the refractive index of the medium in which P is placed.

When a thin centric pencil is considered, K must be very close to A, and in such circumstances the distances PK and QK may be regarded as equal to PA and QA, *i.e.* to p and q. We then have for a thin centric pencil

$$\frac{p-r}{q-r} = \frac{\mu'}{\mu_0} \cdot \frac{p}{q}$$

or

$$\mu_0 pq - \mu_0 qr = \mu' pq - \mu' pr$$

On dividing by pqr we have

$$\frac{\mu_0}{r} - \frac{\mu_0}{p} = \frac{\mu'}{r} - \frac{\mu'}{q}$$

or

$$\frac{\mu'}{q} - \frac{\mu_0}{p} = \frac{\mu' - \mu_0}{r}$$

If now the figure be supposed to rotate round the axis PCA, the extreme ray PK will trace out the limits of the incident cone from P, and it is clear that all the constituent rays of this cone will after refraction proceed as if diverging from the point Q; in other words, Q is the virtual image of P. Indeed, the image of a real object placed in any position before the concave surface of a dense medium is always virtual.

When the first medium is air, $\mu_0 = 1$, and we get the ordinary formula of the books :—

$$\frac{\mu}{q} - \frac{1}{p} = \frac{\mu - 1}{r}$$

This, however, is neither its simplest nor its most easily remembered form, but before proceeding further we will show that this formula applies equally to convex surfaces. A mathematician would know, from the method of proof, that the result must be universally true however the signs of p, q, and r were changed, but we feel that this statement will not be convincing to all our readers.

Convex Surface of Dense Medium.—As before, let PA $= p$, QA $= q$, CA $= r$ (Fig. 25), and consider the incident ray PK having NKP for its angle of incidence, or ϕ; if then NKQ be the angle of refraction, or ϕ',

$$\frac{\mu'}{\mu_0} = \frac{\sin \phi}{\sin \phi'} = \frac{\sin \text{NKP}}{\sin \text{NKQ}} = \frac{\sin \text{PKC}}{\sin \text{QKC}}$$

Now, in the triangle PKC, $\frac{\text{PC}}{\text{CK}} = \frac{\sin \text{PKC}}{\sin \text{KPC}}$

and in the triangle QKC, $\frac{\text{QC}}{\text{CK}} = \frac{\sin \text{QKC}}{\sin \text{KQC}}$

[A warning must here be given about signs: note that PC is measured in the opposite direction to CK, so the angle

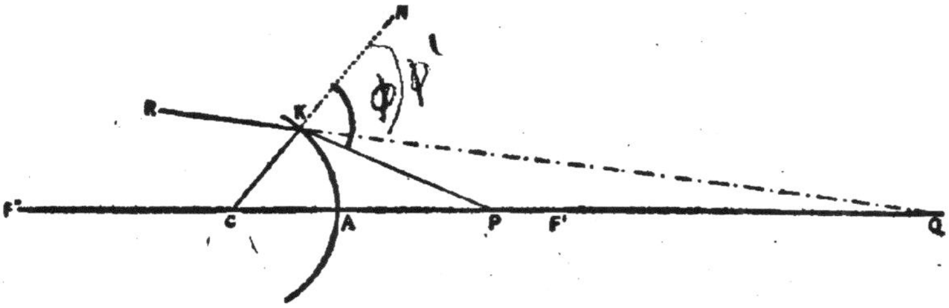

FIG. 25.

PKC is measured in the opposite direction to KPC, one being clockwise and the other being counter-clockwise, and a similar precaution is necessary about the angles QKC and KQC. Want of attention to points like these may give an entirely erroneous formula.]

On dividing the first expression by the second, we get

$$\frac{\text{PC}}{\text{QC}} = \frac{\sin \phi}{\sin \phi'} \cdot \frac{\sin \text{KQC}}{\sin \text{KPC}} = \frac{\mu'}{\mu_0} \cdot \frac{\sin \text{KQP}}{\sin \text{QPK}} = \frac{\mu'}{\mu_0} \cdot \frac{\text{PK}}{\text{QK}}$$

i.e. $\frac{\text{PA} - \text{CA}}{\text{QA} - \text{CA}}$ or $\frac{p - r}{q - r} = \frac{\mu'}{\mu_0} \cdot \frac{\text{PK}}{\text{QK}}$

Consequently, in the limiting case when K is very near to A—

$$\frac{p-r}{q-r} = \frac{\mu'}{\mu_0} \cdot \frac{p}{q}$$

or

$$\frac{\mu'}{q} - \frac{\mu_0}{p} = \frac{\mu' - \mu_0}{r}$$

The formula for a convex surface is therefore identical with that for a concave surface, but whereas the latter makes incident rays more divergent (unless $p < r$), a convex surface always makes them more convergent, and this depends on the fact that in the case of the concave surface CA or r is measured in the same direction as PA, and in the case of the convex surface in the reverse direction.

It will be noted that in both Fig. 24 and in Fig. 25 the image Q is virtual, and hence P and Q are not conjugate in the sense defined on p. 20. It would be found, for instance, that if in Fig. 24 the object were placed at Q the image would be virtual and situated somewhere between Q and C. As stated before, P and Q are only interchangeable or conjugate when the image Q is real.

In Fig. 25 we expect that if PA were increased a little so as to render the incident cone less divergent, the refracted ray KR would be parallel to the axis PAC. Now, if KR were parallel to the axis, QA would be infinite, so let us put $q = \infty$ in our formula.

Then $$\frac{\mu'}{\infty} - \frac{\mu_0}{p'} = \frac{\mu' - \mu_0}{r} \quad \text{or} \quad p' = \frac{-\mu_0 r}{\mu' - \mu_0}$$

We see, therefore, that when the source of light is put at a distance p' the emergent rays are parallel. This position is called the First Principal Focus, and is denoted by F′, while the distance F′A is denoted by f' instead of p'. We have then $f' = \dfrac{-\mu_0 r}{\mu' - \mu_0}$.

Now suppose that the incident rays are parallel, or that p is infinite,

$$\frac{\mu'}{q'} - \frac{\mu_0}{\infty} = \frac{\mu' - \mu_0}{r} \quad \text{or} \quad q' = \frac{\mu' r}{\mu' - \mu_0}$$

The point at which incident parallel rays come to a focus is called the Second Principal Focus and is represented by F″ in the figure, while the distance F″A is commonly denoted by f''; so, replacing q' by its technical symbol, we have $f'' = \dfrac{\mu' r}{\mu' - \mu_0}$.

It should be noted that f'' has the same sign as r, and indeed F″ is seen to be on the same side of the surface as C; F′, however, is on the opposite side, as is obvious from the expression for f' being preceded by a negative sign.

Now, since the effect of a concave surface is to render incident rays more divergent (unless $p < r$), it will be clear,

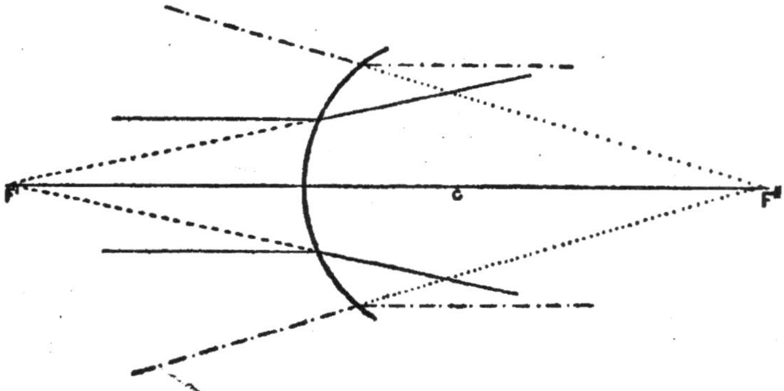

FIG. 26.

on considering refraction at such a surface, that in order that rays may emerge parallel they must have been initially converging towards the point F′ (Fig. 26). Consequently the First Principal Focus F′ in this case is virtual. Again, the Second Principal Focus F″ is also virtual, as incident parallel rays (the spaced and dotted lines) after refraction diverge as if they were proceeding from F″. It will be noticed that here also f'' has the same sign as r, while f' has the opposite sign.

Since $\dfrac{\mu' r}{\mu' - \mu_0} - r = \dfrac{\mu_0 r}{\mu' - \mu_0}$, $\quad f'' - r = -f'$

and we see, when the vertex of the spherical surface is denoted by A, that both in Fig. 25 and in Fig. 26

$$F''A - CA = AF' \text{ or } -F'A, \text{ or } f'' - r = -f'.$$

Note that both F′ and F″ are real when the dense medium presents a convex surface.

Again, it is clear that

$$f'' = -\frac{\mu'}{\mu_0}f' \text{ for } \frac{\mu' r}{\mu' - \mu_0} = -\frac{\mu'}{\mu_0}\left(-\frac{\mu_0 r}{\mu' - \mu_0}\right)$$

We can now express our formula in its simplest form, for on dividing—

$$\frac{\mu'}{q} - \frac{\mu_0}{p} = \frac{\mu' - \mu_0}{r} \quad \text{by} \quad \frac{\mu' - \mu_0}{r} \text{ we get}$$

$$\frac{f'}{p} + \frac{f''}{q} = 1$$

This form, $\frac{f'}{p} + \frac{f''}{q} = 1$, is analogous to the expression we have already found for reflection at spherical surfaces; then, indeed, there was no difference between the positions of the first and second principal foci, as they both coincided in one point. It will be found that $\frac{f'}{p} + \frac{f''}{q} = 1$ is much the easiest form to remember, especially as we shall find that exactly the same formula holds good with lenses.

Ex. (1).—The refractive condition of the human eye may be very closely approximated by a convex spherical segment of radius -5 mm. bounding a medium of index $\frac{4}{3}$. Where will its focus be for incident parallel rays, and what will be the position of its First Principal Focus?

Here the first medium is air, so $\mu_0 = 1$ and $\mu' = \frac{4}{3}$,

$$f'' = \frac{\mu' r}{\mu' - \mu_0} = \frac{\frac{4}{3}(-5)}{\frac{4}{3} - 1} = -20 \text{ mm.}$$

and $$f' = -\frac{\mu_0}{\mu'}f'' = -\tfrac{3}{4}(-20) = 15 \text{ mm.}$$

So F′ is situated 15 mm. in front of the cornea, and F″ 20 mm. behind it.

If the focussing mechanism were not called into play, where would the image be formed by such an eye of an object 16·5 cm. distant?

$$\frac{f'}{p}+\frac{f''}{q}=1 \quad \therefore \frac{f''}{q}=\frac{p-f'}{p} \quad \text{or} \quad q=\frac{pf''}{p-f'}$$

Here $p = 165$ mm.

$$\therefore q=\frac{165 \times -20}{165-15}=\frac{-330}{15}=-22 \text{ mm.}$$

Presuming the curvature of the cornea and the other conditions of the eye to be normal, we should infer that if it could not see an object distinctly at a greater distance than 16·5 cm., it must be 2 mm. longer than normal. We have just found that a human eye that can see very distant objects like stars distinctly must have its retina 20 mm. behind the cornea.

Such problems are delightfully simple and easy, but mistakes are frequently made when the object is in the dense medium. By attending to the rule that μ_0 is used to indicate the index of that medium in which the object is situated, all difficulty will be avoided. We will give an example of the way such questions should be treated.

Ex. (2).—There is a speck within a spherical glass ball ($r = 4$ cm.) distant 2 cm. from one surface. Where will its image be formed as seen from either side when $\mu = 1·5$?

The first point to remember is that the object will form an image, whether real or virtual, irrespective of the presence or absence of an eye to see it. Consequently, neglect the position of the observer, except in so far as it determines which surface of the sphere is being considered, and use the "Standard Notation" for signs.

In this case the object P is in the dense medium, so $\mu_0 = 1·5$, and the final medium is air, so $\mu' = 1$. Suppose the ball to be so placed that the surface nearest to P is to the left; call this side A, and the distal side B.

A. Then PA or $p = -2$, and since $r_1 = -4$,

$$f_1' = \frac{-\mu_0 r_1}{\mu' - \mu_0} = \frac{6}{-0\cdot5} = -12$$

and
$$f_1'' \text{ or } -\frac{\mu'}{\mu_0} f_1' = \frac{12}{1\cdot5} = 8$$

$$\therefore q \text{ or } \frac{pf_1''}{p - f_1'} = \frac{-16}{10} = -1\cdot6 \text{ cm.}$$

The image is virtual, within the glass, 1·6 cm. from A.

B. Here p' or PB $= 6$, and $r_2 = 4$;

$$\text{so now } f_2' \text{ or } \frac{-\mu_0 r_2}{\mu' - \mu_0} = 12, \text{ while } f_2'' = -8$$

$$\therefore q' \text{ or } \frac{p' f_2''}{p' - f_2'} = \frac{-48}{-6} = +8 \text{ cm.}$$

The image is virtual, and is formed at the distance of the diameter from B, that is at the surface A.

In this case, as the second medium is to be regarded as the refracting medium, it is obvious that at the surface B the air presents a concave surface, and the incident divergent cone from P is rendered less divergent, for QB is greater than PB. Indeed, in every case (except when $p < r$, as in case A) the effect of refraction at the concave surface of a rare medium is to increase the convergence or diminish the divergence. This can easily be remembered by considering the case of a biconvex lens: convergence occurs at the first surface, of course, because refraction occurs at the convex surface of a dense medium; and this convergence is increased by the refraction at the second surface, which presents the concave surface of a rare medium (air). When the rare medium presents a convex surface, divergence results without any exception.

Geometrical Construction of the Image.—The construction is almost exactly the same as that we employed before, when dealing with a spherical mirror. The only difference is due to the fact that now F′ and F″ no longer coincide, but are situated on opposite sides of the refracting surface.

As before, we give two methods to find the image of a point not on the principal axis, and one method for a point that is on the principal axis. As we are assuming that only centric pencils contribute to the formation of the image, we

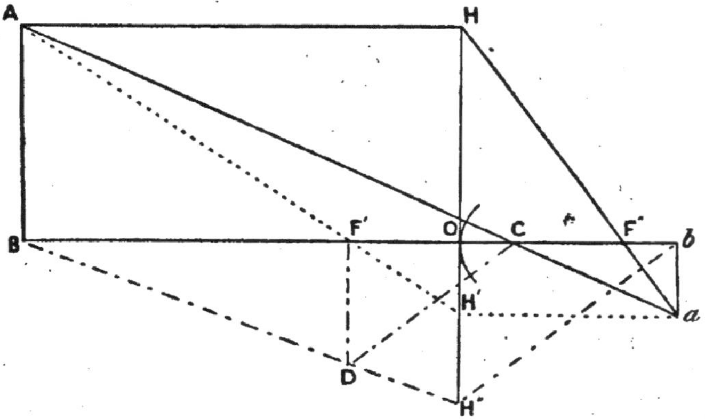

Fig. 27.

draw the principal plane HOH″ tangential to the spherical surface at its vertex O (Fig. 27). AB represents the object, BOC the principal axis, F′ and F″ the first and second foci of the spherical refracting surface whose centre is C.

(A) *Point not on the Principal Axis.*—

(1) Draw ACa through the centre C. Through A draw a line parallel to the principal axis, meeting the principal plane at H. Draw HF″a through F″ to meet the line ACa in a.

Then a, the point of intersection of HF″ and AC produced, marks the position of the image of the point A.

(2) (Dotted lines.) Draw ACa as before through the centre C; from A draw a line through the first focus F′ to meet the principal plane at H′. Draw

H′a parallel to the principal axis until it meets ACa in a.

The point a is the image of the point A.

It would appear, then, that the incident diverging cone HAH′ becomes, after refraction, the converging cone HaH′. This is not strictly true, for the line AH does not represent any ray that is actually incident on the refracting surface; but we are justified in asserting that a small centric pencil from A will come to its conjugate focus at the point of intersection of HF″ and AC produced. Again, we know that all rays that pass through F′, such as AF′H′, must, after refraction, proceed in a direction (H′a) parallel to the axis.

(B) *Point on the Principal Axis.*—(Spaced and dotted lines.)

Through B draw any line BDH″ cutting the first focal plane in D and the principal plane at H″; join DC and draw H″b parallel to DC, cutting the principal axis in b.

Then the point b is the image of the point B.

The reason of this construction is obvious from the property of the focal planes described on p. 26. Light from any point on this plane will, after refraction, travel in rays parallel to that axis on which the point lies. Now, DC is, of course, the axis on which D lies, and consequently the incident light ray DH″ must, after refraction, take the direction H″b parallel to DC.

Size of the Image.—In just the same way as we found the height of the image formed by reflection at a mirror, we can find $\frac{i}{o}$ in this case.

(1) Noting that ba is equal to OH′, we find, by similar triangles, that—

$$\frac{i}{o}=\frac{ba}{BA}=\frac{OH'}{BA}=\frac{F'O}{F'B}=\frac{F'O}{F'O-BO}=\frac{f'}{f'-p}$$

(2) And as BA is equal to OH,

$$\frac{i}{o}=\frac{ba}{BA}=\frac{ba}{OH}=\frac{F''b}{F''O}=\frac{F''O-bO}{F''O}=\frac{f''-q}{f''}$$

Exactly the same construction is used when the refracting surface is concave, in which case F″ with C would lie on the object side of the refracting surface, and F′ would lie on the opposite side. It should be noted, also, that when the refracting medium is dense and presents a concave surface the image is erect and virtual, but when the surface is convex the image is inverted and real as long as p is greater than f'; if $p = f'$ the refracted rays are parallel, and if $p < f'$ they are divergent. We will give one or two examples where $\mu_0 = 1$ (the first medium), and $\mu' = \frac{4}{3}$.

Ex. (1).—A refracting surface presents a convex surface of radius -4 cm., and at a distance of 8 cm. in front of it is placed an object 5 cm. high. Give the position, the character, and the height of the image.

We must first find the foci.

$$f' = \frac{-\mu_0 r}{\mu' - \mu_0} = -\frac{-4}{\frac{4}{3} - 1} = 12 \text{ cm.}$$

and $$f'' = -\frac{\mu'}{\mu_0} f' = -\tfrac{4}{3}(12) = -16 \text{ cm.}$$

then $$q = \frac{pf''}{p - f'} = \frac{8 \times (-16)}{8 - 12} = 32 \text{ cm.}$$

The image is therefore formed 32 cm. on the object side of the refracting surface, so it is virtual.

$$\frac{i}{o} = \frac{f'}{f' - p} = \frac{12}{12 - 8} = 3$$

or if we use the second formula—

$$\frac{i}{o} = \frac{f'' - q}{f''} = \frac{-16 - 32}{-16} = 3$$

The same result must be obtained in either case: the image is erect because $\frac{i}{o}$ is positive, and three times higher than the object. Consequently, the height of the image is 15 cm.

This is a case in which p is less than f', and the refracted rays diverge, though less than the original incident rays, so a virtual image is formed at a greater distance than that of the object.

When speaking of the effect of concave refracting surfaces, we said that incident rays were always made more divergent unless p were less than r. As an example, we will now take a case where p is less than r.

(2) An object 5 cm. high is placed 3 cm. in front of a concave refracting surface of radius 4 cm. Where will the image be formed, and what will be its height?

In this case, as the surface is concave r is positive; so

$$f' = -\frac{\mu_0 r}{\mu' - \mu_0} = -\frac{4}{\frac{4}{3} - 1} = -12 \text{ cm. and } f'' = -\tfrac{4}{3}f' = 16 \text{ cm.}$$

$$q = \frac{pf''}{p - f'} = \frac{3 \times 16}{3 + 12} = \frac{16}{5} = 3{\cdot}2 \text{ cm.}$$

The image is therefore situated 3·2 cm. in front of the surface, or 2 mm. further off than the object; in other words, the divergent cone from each point of the object is made rather less divergent,

and $$\frac{i}{o} = \frac{f'}{f' - p} = \frac{-12}{-12 - 3} = \frac{4}{5}$$

So the image is virtual and erect, and as the object is 5 cm. high, the height of the image is 4 cm.

Graphic Method for Refraction at a Spherical Surface.—The graphic method that we employed for spherical mirrors can also be very conveniently used in this case. As an example we give in Fig. 28 the method as applied to the normal human eye. The complex system of the eye with its cornea and lens will be shown on p. 102 to be almost exactly equivalent, from an optical standpoint, to a single refracting convex spherical surface of radius $-5{\cdot}25$ mm., with f' or F′H equal to $+15{\cdot}54$ mm. and f'' or F″H equal to $-20{\cdot}79$ mm.

On the horizontal straight line PH we mark off the point F′ so that F′H = 15·54 mm., and from F′ we raise a vertical line such that F′F″ = $-20{\cdot}79$ mm., for we consider lines measured from left to right and from below upwards as positive, and those in the reverse directions as negative. Where will the image be formed by such a refracting system of an object $8\frac{1}{4}$ cm. distant?

Make PH = $8\frac{1}{4}$ cm., and join PF″ and produce it to cut HQ in Q. Measure QH; it is found to be −25·56 mm. The image is therefore formed 25·56 mm. behind H, which denotes the cornea. Now, the "standard" eye that can see

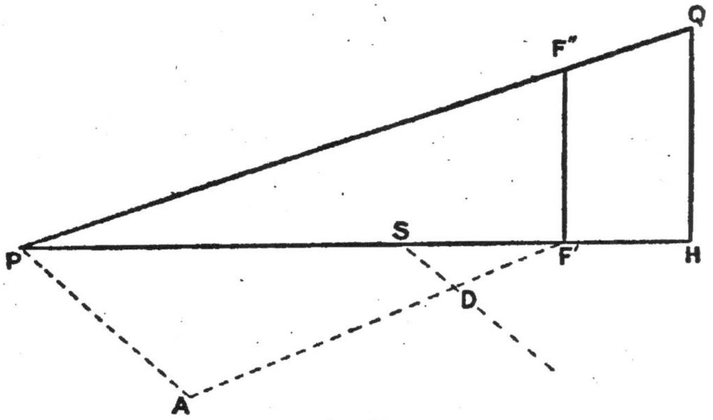

FIG. 28.

very distant objects distinctly must have its retina at a distance of 20·79 mm. from its cornea; suppose that a certain person could not see objects further off than 82·5 mm. (this case), then his retina must be 25·56 mm. from his cornea, or his eye must be 4·77 mm. too long, if his eye were otherwise normal.

The size of the retinal image is given by our previous formula, $\frac{i}{o} = \frac{F'H}{F'P} = \frac{15 \cdot 54}{-66 \cdot 96}$, so that the image is inverted and real, and a little less than one-fourth of that of the object.

Further, our diagram will tell us what glass would correct this eye for distance. Spectacles should always be worn in the first focal plane of the eye, *i.e.* about half an inch in front of the cornea, so that F′ marks the position of the correcting glass, and it must be of such strength that parallel rays from a distant object should form their image at P. This means that the second focal distance (f'') of the correcting glass for distance must be equal to PF′, which is found by measurement to be 67 mm. approximately. Now, a lens of which $f'' = 67$ mm. is one of power $\frac{-1000}{67}$, or nearly -15D, as will be shown in the next chapter. Indeed, if preferred, this division sum may be done graphically in this way. Draw a line F′A -50 mm. long in any direction that makes an acute angle with F′P (Fig. 28); mark off S on PF′, making SF′ equal to 20 mm. Join AP and draw SD parallel to AP; then F′D in millimetres gives the power of the correcting glass in dioptres (see p. 80). For of course

$$\frac{F'D}{SF'} = \frac{F'A}{PF'}, \quad \textit{i.e.}\ \frac{F'D}{20} = \frac{-50}{67} \quad \therefore F'D = \frac{-1000}{67} \approx -15$$

Note that both F′D and F′A are measured in the negative direction.

If the eye were 3·23 mm. too short, where should the object be to form a distinct image on the retina?

Make QH $-17{\cdot}56$ mm. (*i.e.* $-20{\cdot}79 + 3{\cdot}23$). Join F″Q, and produce to meet the base line in P (Fig. 29). If accurately drawn, P will lie 84·46 mm. to the right of H. This means that PH is negative; thus, unless very great focusing power were used, the eye would not be able to see any real object distinctly, for light would have to converge as if towards a point 84·46 mm. behind the eye in order to come to a focus on the retina. It is unnecessary to determine the size of the image, as we have seen that no real object could be seen. The correcting glass must have f'' equal to PF′, which by measurement we find to be -100 mm.

$$\therefore D = \frac{-1000}{-100} \text{ or } +10$$

Or if we wish to do this division graphically, we draw F′A 50 mm. long in any direction, making an acute angle with

F′P. We then mark off S between P and F′, making SF′ equal to −20 mm. Then by joining PA and drawing SD parallel to it, cutting F′A in D, we measure F′D. We find

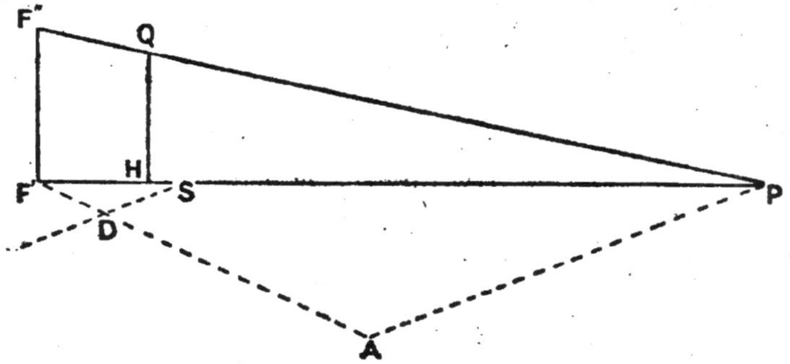

FIG. 29.

that F′D = 10 mm., so his correcting glass is +10D. The graphic method is given in this case to show the generality of the method, though here there is no special advantage in it. As before, we have—

$$\frac{F'D}{SF'} = \frac{F'A}{PF'}, \ i.e. \ \frac{F'D}{-20} = \frac{50}{-100} \quad \therefore F'D = \frac{-1000}{-100} = 10$$

***Eccentric Pencils—Focal Lines.**—In the preceding sections we have been considering the refraction of thin centric pencils, of which the central ray traversed the centre of the refracting surface. We will now take the case of a thin eccentric pencil that does not pass through the centre.

Let O be a luminous point on the axis of the concave surface whose centre is at C (Fig. 30). The thin pencil POQ will be refracted in the direction RR′, so that the refracted rays, if produced backwards, will meet at F_1, and will cut the axis in a line in the neighbourhood of F_2, so that the refracted pencil will be astigmatic. If the incident pencil were of a pyramidal shape, the prolongations of the refracted rays would form a figure something like Fig. 15, only in the case shown in Fig. 30 F_1P is greater than F_2P. A similar sphenoid will be formed between the secondary and primary focal lines, but the lateral surfaces of the refracted beam will now intersect in the secondary focal line at F_2 before the final edge of the sphenoid is formed at F_1, the primary focal line. If OP were greater than $\mu(F_1P)$, F_2P would be greater than F_1P (see Appendix, p. 116); in such a case the refracted beam would take exactly the form represented in Fig. 15.

The formulæ for the lengths F_1P or v_1, and for F_2P or v_2, when the source of light O is at a distance OP or u, will be merely stated in this place—

$$\frac{\mu \cos^2 \phi'}{v_1} - \frac{\cos^2 \phi}{u} = \frac{\mu}{v_2} - \frac{1}{u} = \frac{\sin(\phi - \phi')}{r \sin \phi'}$$

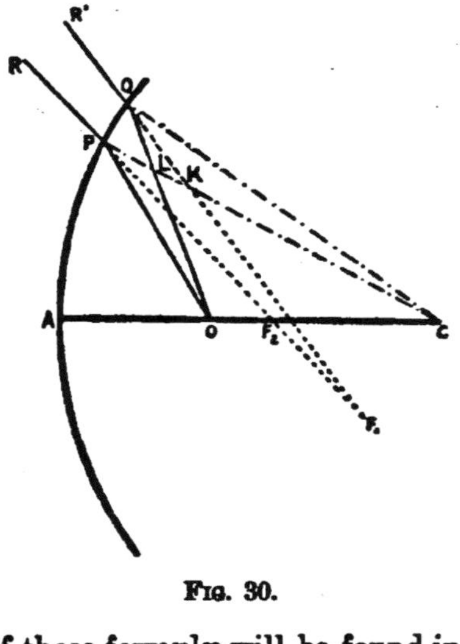

FIG. 30.

The proof of these formulæ will be found in the Appendix, pp. 114–116.

Conversion of Formulæ of Refraction into those of Reflection.—On p. 35 it was stated that $\frac{\sin \phi}{\sin \phi'} \left(\text{or } \frac{\mu'}{\mu_0}\right) = \frac{V_1}{V_2}$, *i.e.* the ratio of the speed of light in the first medium to its speed in the second medium. Now, when reflection occurs at a surface, it is clear that the speed of the light is unchanged in absolute amount, for it is in the same medium, but its direction is reversed; in other words, the sign is changed, so that for reflection we must consider V_2 as equal to $-V_1$. We have therefore $\frac{\mu'}{\mu_0}$ or $\frac{V_1}{V_2} = \frac{V_1}{-V_1} = -1$, so that every formula applying to refraction must apply to the similar case when reflection is considered, if we simply replace the expression for the relative index of refraction $\frac{\mu'}{\mu_0}$ by -1. Indeed, this is one of the best tests we have to determine whether a formula for refraction is correct or not. Replace μ' by -1 and μ_0 by 1, and see if the correct formula for reflection is given. We will give a few examples to illustrate the generality of this method of conversion.

Refraction.	Reflection.
$\frac{\sin\phi}{\sin\phi'} = \frac{\mu'}{\mu_0}$	$\frac{\sin\phi}{\sin\phi'} = -1 \quad \therefore \phi = -\phi'$
$\frac{\mu'}{q} - \frac{\mu_0}{p} = \frac{\mu' - \mu_0}{r}$	$\frac{-1}{q} - \frac{1}{p} = \frac{-2}{r}$ or $\frac{1}{p} + \frac{1}{q} = \frac{2}{r}$
$\frac{\mu}{v_2} - \frac{1}{u} = \frac{\sin(\phi - \phi')}{r\sin\phi'}$	$\frac{-1}{v_2} - \frac{1}{u} = \frac{\sin 2\phi}{-r\sin\phi} = \frac{2\sin\phi\cos\phi}{-r\sin\phi}$ or $\frac{1}{v_2} + \frac{1}{u} = \frac{2\cos\phi}{r}$
$\frac{\mu\cos^2\phi'}{v_1} - \frac{\cos^2\phi}{u} = \frac{\sin(\phi - \phi')}{r\sin\phi'}$	$\frac{-\cos^2\phi}{v_1} - \frac{\cos^2\phi}{u} = \frac{2\sin\phi\cos\phi}{-r\sin\phi}$ or $\frac{1}{v_1} + \frac{1}{u} = \frac{2}{r\cos\phi}$

The reader need not therefore burden his memory with any of the special formulæ for eccentric reflection, as the conversion of the formulæ for eccentric refraction is so easy. As another aid to memory, I may add that any formula for a curved surface becomes true for a plane surface if we make $r = \infty$, for this is equivalent to making the curvature 0. For instance, taking our last illustrations for eccentric pencils—

$$\frac{\mu\cos^2\phi'}{v_1} - \frac{\cos^2\phi}{u} = \frac{\mu}{v_2} - \frac{1}{u} = \frac{\sin(\phi - \phi')}{r\sin\phi'}$$

becomes, when r is made infinite—

$$\frac{\mu\cos^2\phi'}{v_1} - \frac{\cos^2\phi}{u} = \frac{\mu}{v_2} - \frac{1}{u} = 0$$

$$\text{or} \quad v_1 = \mu u\frac{\cos^2\phi'}{\cos^2\phi} \quad \text{and} \quad v_2 = \mu u$$

These are the distances of the primary and secondary focal lines when a thin oblique pencil undergoes refraction at a plane surface, μ denoting the relative index of refraction (Appendix, pp. 111–113).

QUESTIONS.

(1) Assuming that the human eye is a simple refracting system, of which the first focal distance is 15 mm. and the second focal distance −20 mm., where would the image be formed, and what would be its height, if an object 6 mm. high were at a distance of 150 mm.?

(2) What curvature must be given to the refracting surface when $\mu = \frac{4}{3}$ in order that the previous object at a distance of 150 mm. may form a real image at a distance of 20 mm.? What would be its height? Compare the size of the retinal images in the axial myopia (the first case in which the eye is long enough to form a distinct image of the object) and the curvature myopia (the second case).

(3) If the back of a glass sphere ($\mu = \frac{3}{2}$) be silvered, where will be the image that is formed by one reflection and one refraction of a speck that is halfway between the centre and the silvered side?

CHAPTER VI.

LENSES.

ANY refracting medium bounded by two curved surfaces which form arcs of spheres is known as a spherical lens; the axis of the lens is the line joining the centres of the spheres, and that part of the axis lying between the two surfaces gives the thickness of the lens.

Thus in Fig. 31 the surface A is the arc of a sphere that has its centre at C_1, and the surface B is the arc of another

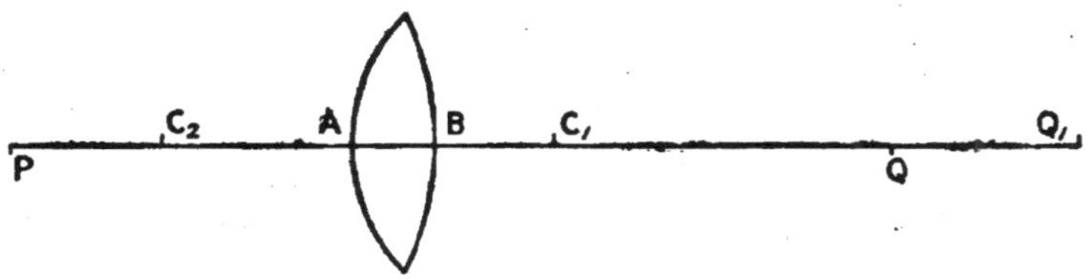

FIG. 31.

sphere that has its centre at C_2: the axis of the lens is the line joining C_1 and C_2, and the thickness of the lens is the length of the line AB. These are the only definitions we shall require for the present. It will be apparent that if one of the surfaces be plane, it may be regarded as a spherical surface of which the radius is infinite.

Thin Lenses. Thin Axial Pencils.—The conjugate focal distances of an axial pencil of a thin lens can be very easily determined from the formula that we have already found for a single spherical surface. It should be noted that the term "centric" is no longer equivalent to the term "axial," as an oblique pencil that passed through the centre of one spherical surface would not pass through the centre of the other surface unless they coincided (see p. 84). We shall eventually determine the position of what is called the Optical Centre of a lens, and discuss its properties; at present we are only concerned with axial pencils.

Let P (Fig. 31) be a luminous point on the axis of the lens, and let Q_1 be the image of P due to refraction at the first surface A, and let $\frac{\mu'}{\mu_0}$ be the relative refractive index between the first medium and the second medium (*e.g.* air and glass). When $p = \text{PA}$, $q_1 = \text{Q}_1\text{A}$, and when r_1 is the radius of the first surface,

$$\frac{\mu'}{q_1} - \frac{\mu_0}{p} = \frac{\mu' - \mu_0}{r_1} \quad . \quad . \quad . \quad . \quad . \quad (a)$$

The light from P, after traversing the first surface A, will be travelling in the direction towards Q_1. On emerging at the second surface into the original medium (μ_0), refraction again takes place, and the relative index is now $\frac{\mu_0}{\mu'}$. As the lens is considered to be of negligible thickness, Q_1B may be regarded as equal to Q_1A or q_1.

Regarding, then, Q_1 as the source for the second refraction at B, of which the radius is r_2, and putting q for QB, the distance of the final image, we have

$$\frac{\mu_0}{q} - \frac{\mu'}{q_1} = \frac{\mu_0 - \mu'}{r_2}$$

On adding (*a*)

$$\frac{\mu'}{q_1} - \frac{\mu_0}{p} = \frac{\mu' - \mu_0}{r_1}$$

we get

$$\frac{\mu_0}{q} - \frac{\mu_0}{p} = (\mu' - \mu_0)\left(\frac{1}{r_1} - \frac{1}{r_2}\right) \quad . \quad . \quad (b)$$

This is the standard formula for a thin lens; when the lens is in air (as is almost always the case) $\mu_0 = 1$, and we can suppress the dash and write

$$\frac{1}{q} - \frac{1}{p} = \overline{\mu - 1}\left(\frac{1}{r_1} - \frac{1}{r_2}\right) \quad . \quad . \quad . \quad . \quad (c)$$

To find the First Principal Focus (F′), as before, we make

the emergent rays parallel, or we make q infinite; we then have $\frac{1}{q} = 0$, and writing f' for this special value of p, we get

$$\frac{1}{f'} = -\overline{\mu - 1}\left(\frac{1}{r_1} - \frac{1}{r_2}\right)$$

To find the Second Principal Focus (F″), or the focus for incident parallel rays, we make p infinite; so, substituting f'' for this special value of q, we get

$$\frac{1}{f''} = \overline{\mu - 1}\left(\frac{1}{r_1} - \frac{1}{r_2}\right)$$

We see that $f' = -f''$, which is always the case when the initial and the final media are the same; so, on substituting these values in (c), we find that

$$\frac{1}{q} - \frac{1}{p} = \frac{1}{f''} \quad \text{or} \quad \frac{f''}{q} - \frac{f''}{p} = 1$$

and finally our old formula

$$\frac{f'}{p} + \frac{f''}{q} = 1 \quad . \quad . \quad . \quad . \quad . \quad . \quad \text{(A)}$$

[It may be here noted that when thick lenses are considered, if the thickness be denoted by t, the term $\frac{-(\mu - 1)^2 t}{\mu r_1 r_2}$ must be added to the expressions for $\frac{1}{f''}$ and subtracted from that for $\frac{1}{f'}$.

Thus $$\frac{1}{f''} = \overline{\mu - 1}\left(\frac{1}{r_1} - \frac{1}{r_2} - \frac{\overline{\mu - 1}t}{\mu r_1 r_2}\right)$$

and $$\frac{1}{f'} = -\overline{\mu - 1}\left(\frac{1}{r_1} - \frac{1}{r_2} - \frac{\overline{\mu - 1}t}{\mu r_1 r_2}\right)$$

or $$\frac{1}{f''} = -\frac{1}{f'} = \frac{\mu - 1}{r_1 r_2}\left(r_2 - r_1 - \frac{\overline{\mu - 1}t}{\mu}\right)$$

These expressions give the focal lengths correctly, and for some purposes this is all that is required, but they do not enable us to determine the position of F' and F''. We shall find later (p. 91) that when a thick lens is considered, there are two Principal Points towards which the two focal distances must be measured respectively.]

Let us suppose that the lens in Fig. 31 is of negligible thickness and has for its radii of curvature $r_1 = -2$ cm. and $r_2 = 4$ cm. What will be its focal distances when $\mu = 1{\cdot}5$?

$$\frac{1}{f'} = -\overline{\mu - 1}\left(\frac{1}{r_1} - \frac{1}{r_2}\right) = -\frac{1}{2}\left(\frac{1}{-2} - \frac{1}{4}\right) = \frac{3}{8}$$

$$\therefore f' = \tfrac{8}{3} \text{ cm. and } f'' = -\tfrac{8}{3} \text{ cm.}$$

Consequently, if a source of light were placed at $2\frac{2}{3}$ cm. distance from such a lens, the emergent rays would be parallel. If, however, the incident rays were parallel, as for instance from the sun, they would converge to a focus $2\frac{2}{3}$ cm. on the far side of it, as the negative sign shows that the Second Principal Focus is behind the lens.

Ex. (1) If an object were placed 24 cm. in front of such a lens, where would the image be formed?

Here $p = 24$, and we wish to find the corresponding value of q.

$$\frac{f'}{p} + \frac{f''}{q} = 1 \qquad \therefore \frac{f''}{q} = \frac{p - f'}{p} \quad \text{or} \quad q = \frac{pf''}{p - f'}$$

so $$q = \frac{24(-\frac{8}{3})}{24 - \frac{8}{3}} = -\frac{24 \times 8}{64} = -3 \text{ cm.}$$

The image would therefore be real, and it would be formed 3 cm. behind the lens.

(2) An object is placed 24 cm. in front of a concave lens, of which $r_1 = 2$ cm. and $r_2 = -4$ cm. ($\mu = 1{\cdot}5$). Where will its foci be situated, and where will the image of the object be formed?

This lens has the same curvature as that in (1), but in the reversed direction, so f' will be found to be $-\frac{8}{3}$ cm. and $f'' = \frac{8}{3}$ cm.

Hence $$q \text{ or } \frac{pf''}{p - f'} = \frac{24(\frac{8}{3})}{24 + \frac{8}{3}} = 2{\cdot}4 \text{ cm.}$$

The image in this case is virtual, and is formed 2·4 cm. in front of the lens.

It should be noted that in all converging lenses f' is positive (*i.e.* F′ is on the same side of the lens as the object P) and f'' negative, and they both are real, whereas in all diverging lenses f' is negative (*i.e.* F′ is on the side opposite to P) and f'' positive, and they both are virtual. Moreover, all converging lenses are thickest in the middle, whereas all lenses which are thinnest in the middle are diverging in function. The reverse of these statements is not true, although it is frequently alleged to be so; *e.g.* a lens can be constructed that is thickest in the middle and yet be diverging in function.

When a lens is converging in function, the image is real and inverted (Fig. 32) or virtual and erect (Fig. 33), according as the distance of the object is greater or less than the first focal distance of the lens. With diverging lenses the image of a real object is always virtual.

In Figs. 32 and 33 the object is represented to the right of the lens, and therefore F′ must be on the same (the incident) side, and as the lenses are of the same focal length ($f' = 4$ cm.) we make OF′ in each case equal to 4 cm. In Fig. 32 the object AB is placed 14 cm. from the lens, so the image *ab* will be real and inverted, and it will be situated at a distance of 5·6 cm. on the other side of the lens,

for $$q = \frac{pf''}{p - f'} = \frac{14(-4)}{14 - 4} = -5{\cdot}6 \text{ cm.}$$

In Fig. 33 AB is placed 3 cm. from the lens, and the image *ab* is therefore virtual and erect, and it is 12 cm. from the lens on the object side.

$$\text{for } q = \frac{pf''}{p - f'} = \frac{3(-4)}{3 - 4} = 12 \text{ cm.}$$

Geometrical Construction and Size of the Image.—The image is drawn by the same method that was employed in constructing the image formed by refraction at a single spherical surface (Fig. 27, p. 58). Three alterations are, however, necessary: (1) the principal plane HOH′ must be drawn through the centre of the thin lens instead of tangentially to its surface; (2) OF″ must be made equal to F′O; and (3) the point O must be regarded as taking the place of C, which is no longer required (cf. Figs. 32 and 33).

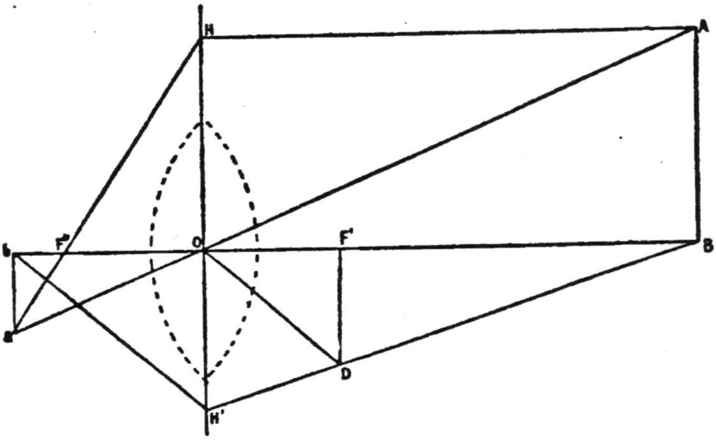

FIG. 32.

When the lens is symmetrical, as in Fig. 32 and 33, O is the optical centre of the lens, and, clearly, any ray of light that traverses O proceeds on its course without deviation, for any refraction that it may undergo on encountering the first surface of the lens will be reversed on emerging from the second surface. It is clear from Fig. 32 that the ray AO*a* cuts the lens at two points such that the tangents at the points of entry and emergence are parallel, so that the light traverses the lens as if it were a plate with parallel sides.

(A) *Point not on the Axis.*—The line AO is drawn and produced to *a*. (1) AH is drawn parallel to the axis, meeting the principal plane in H; HF″ is then drawn through F″ meeting AO*a* in *a* (Fig. 32); or (2) F′A is joined and produced to meet the principal plane in H (Fig. 33), and from H the line H*a* is drawn parallel to the axis, meeting OA*a* in *a*.

(B) *Point on the Axis.*—From B a line BDH′ (Fig. 32) is drawn at any acute angle cutting the first focal plane in D and the principal plane in H′; DO is then drawn, and H′*b* is drawn parallel to DO.

From Fig. 32 we see that—

$$\frac{i}{o}=\frac{ba}{\mathrm{BA}}=\frac{ba}{\mathrm{OH}}=\frac{\mathrm{F}''b}{\mathrm{F}''\mathrm{O}}=\frac{\mathrm{F}''\mathrm{O}-b\mathrm{O}}{\mathrm{F}''\mathrm{O}}=\frac{f''-q}{f''}$$

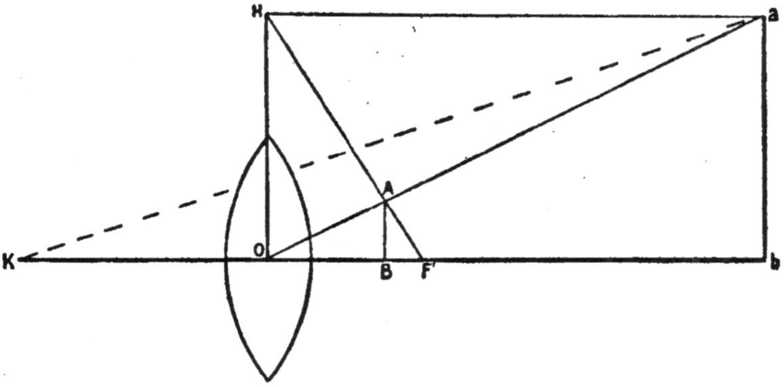

FIG. 33.

and from Fig. 33 that—

$$\frac{i}{o}=\frac{ba}{\mathrm{BA}}=\frac{\mathrm{OH}}{\mathrm{BA}}=\frac{\mathrm{F}'\mathrm{O}}{\mathrm{F}'\mathrm{B}}=\frac{\mathrm{F}'\mathrm{O}}{\mathrm{F}'\mathrm{O}-\mathrm{BO}}=\frac{f'}{f'-p}$$

Therefore when F′O or $f'=4$, BO or $p=14$, and when BA is 8 inches in height, as in Fig. 32,

$$i \text{ or } ba=\frac{f'}{f'-p}\mathrm{BA}=\frac{4}{4-14}8=-3{\cdot}2$$

But if BO or $p=3$ and BA $=2$, as in Fig. 33,

$$i \text{ or } ba=\frac{f'}{f'-p}\mathrm{BA}=\frac{4}{4-3}2=8$$

The formulæ $\frac{f'}{p}+\frac{f''}{q}=1$ and $\frac{i}{o}=\frac{f'}{f'-p}=\frac{f''-q}{f''}$ are thus shown to be universally true for refraction at a single spherical surface, for lenses, and also for reflection, when it is remembered that in the case of a mirror F′ and F″ coincide in one and the same point F.

Figs. 32 and 33 show, however, another expression for $\frac{i}{o}$ that is always true in the case of lenses, for $\frac{ba}{BA}=\frac{bO}{BO}=\frac{q}{p}$. It is, however, rarely necessary to pay any attention to this relation, and it is much easier to remember the few formulæ that are universal, which are given above.

There is, however, one important point to which attention may be particularly directed, as it affords the explanation of certain facts which are obvious enough if it be borne in mind, but puzzling otherwise. It is this, that the angle BOA subtended by the object at the centre of the lens is always equal to the angle *bOa* subtended by the image. The application of this fact may be illustrated by the following example:—the diameter of the sun subtends a visual angle of 31′: what is the diameter of its image as formed by a lens of 1 m. focal length? (tan 31′ = 0·009.)

In this case, of course, an inverted image is formed at F″, and—

$$ba \text{ or } i = f'' \tan 31' = -1000 \times 0{\cdot}009 \text{ mm.} = -9 \text{ mm.}$$

In the case of a mirror, it may be noted that the angle *bOa* is numerically equal to the angle BOA, as may be at once seen by drawing the lines *a*O and AO in Fig. 11. Then if *bOa* be denoted by α, BOA $= -\alpha$, for it is measured in the reverse direction to *bOa*, so that $i = -q \tan \alpha$. When distant objects like the sun or the moon are observed with a reflecting telescope, the incident rays are parallel, and the size of the image formed by the mirror is given by the expression—

$$i = -f \tan \alpha.$$

Ex. (1).—A camera of 6 inches focus shows a distinct image 1 inch high of an object when the ground-glass screen is 6·6 inches away from the lens. At what distance is the object, and what is its height?

Here an inverted image is formed 6·6 inches behind the lens, so

$$q = -6{\cdot}6 \text{ ins.}$$

$$\therefore p \text{ or } \frac{qf'}{q-f''} = \frac{-6{\cdot}6 \times 6}{-6{\cdot}6+6} = 66 \text{ ins., \quad \textit{i.e.} 5 feet 6 ins.}$$

$$\text{and } \frac{i}{o} = \frac{f''-q}{f''} \quad \text{or} \quad \frac{-1}{o} = \frac{-6+6{\cdot}6}{-6} = -\frac{1}{10}$$

$$\therefore o = 10 \text{ ins.}$$

Ex. (2).—The same camera shows an image 1 inch in height of a man who is 6 feet high. How far off is he?

$$\frac{i}{o} = \frac{f'}{f'-p} \quad \text{or} \quad \frac{-1}{72} = \frac{6}{6-p}$$

$$\therefore p - 6 = 6 \times 72 \text{ ins.} \quad \text{or} \quad p = 36\tfrac{1}{2} \text{ feet.}$$

Ex. (3).—If in repairing a bicycle reflex lamp the plane mirror is placed at the focus of the convex lens, will it act in the desired way? No. The light from a distant approaching motor that is incident upon the lens will converge to its focus. From this point on the mirror the reflected light will fall again on the lens, and will emerge as a parallel beam by its previous path. It will therefore only return to the source of light, and hence will give no warning to the driver of the car.

It is for precisely the same reason that the pupil of the eye appears black. It is really red when incident light falls upon it; but the observer necessarily puts his head in the path of the incident light in one direction, so that no light is returned in that direction.

In order to see the back of the eye, an ophthalmoscope is used, which consists essentially of a small mirror with a central perforation, through which the surgeon looks at the eye of the patient. Light is reflected by this mirror into the patient's pupil to the red background of the eye, and returning by its previous path to the mirror is received in part by the surgeon's eye behind the aperture, so that the pupil appears of a bright reddish colour to him. If the retina of the patient be not situated at the focus of his eye, the light on its return from the retina will no longer emerge as a parallel beam, but either as a convergent or divergent pencil, according as the retina is behind or in front of the focus. In such cases, especially when the pupils are widely dilated, they will appear to glow with a ruddy light, even to the unaided eye. For a similar reason, the unsatisfactory reflex lamp may be made efficient by moving the plane mirror a little nearer the lens, so that it is just within its focal distance.

It has been shown that in Fig. 33 the image *ab* is four times the height of the object AB, but this does not necessarily mean that the apparent size of this virtual image, to an eye that perceives it, is four times the apparent size of AB.

Magnification.—The *apparent* size of an object depends upon its distance from the eye that perceives it. If we regard the eye as a simple refracting medium formed of a single spherical surface of radius $-5{\cdot}25$ mm., as on p. 61, the centre of this surface will represent the nodal point (K) of the eye 5·25 mm. behind the cornea. Now, an object BA

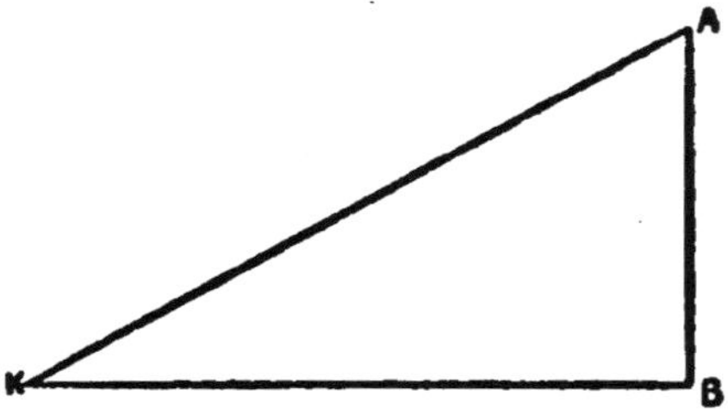

FIG. 34.

(Fig. 34), at a distance KB from the nodal point, will subtend at K the angle BKA or θ. This is called the visual angle. The apparent size of BA is evidently determined by $\tan\theta$ or $\frac{BA}{KB}$. It is obvious that $\tan\theta$ could be increased indefinitely by diminishing KB indefinitely. In other words, the apparent size of an object could be indefinitely increased by bringing the eye indefinitely close to it. The eye, however, is incapable of seeing an object distinctly which lies within a certain distance. The distance KB of this *punctum proximum*, as it is called, from the eye varies in different individuals, and increases with age, so that it is impossible to assign to it any definite value which shall be applicable to all cases. If l denote the least distance for the individual eye considered, then the greatest value that $\tan\theta$ can actually have is by making KB equal to l, when $\tan\theta = \frac{\text{BA}}{l}$ or $\frac{o}{l}$. It is customary to assign an arbitrary value of 10 inches to l.

If, now, a convex glass be placed as in Fig. 33, with the object BA within its first focal distance, and the nodal point of the observer's eye be at K, a virtual image ba will be formed at a distance Kb from the nodal point. Then if θ' is the visual angle subtended by ba at K, $\tan\theta' = \frac{ba}{\text{K}b}$ or $\frac{i}{\text{K}b}$. But we have already found that the maximum size of the object as seen by an unaided eye is given by $\tan\theta$ or $\frac{o}{l}$. The magnification M of a convex glass must therefore be the relation between $\tan\theta'$ and $\tan\theta$.

$$\text{M} = \frac{\tan\theta'}{\tan\theta} = \frac{i}{\text{K}b}\cdot\frac{l}{o} = \frac{l}{\text{K}b}\cdot\frac{f''-q}{f''} = \frac{l}{\text{K}b}\left(1+\frac{q}{f'}\right)$$

When a convex lens is used as a magnifying glass, the image is always virtual, and it therefore lies on the same side of the glass as F′, so the fraction $\frac{q}{f'}$ is always positive $\left(\frac{q}{f'} = \frac{b\text{O}}{\text{F}'\text{O}} = \frac{\text{O}b}{\text{OF}'}\right)$. It is clear from the above expression that M is increased by making Kb as small as possible, and by increasing the value of q. But Kb cannot be made smaller than l (say, 10 ins.), or the image would not be seen distinctly, and as Kb = KO + Ob, say, d + Ob, there is a limit to the value of Ob. The distance d of the lens from the nodal point of the eye cannot be much less than half an inch, so we must make Ob = 9½ inches in order to get the maximum magnifying power out of a lens employed in this way.

In the case above, with a lens of 4 inches focus, with the object so placed as to form a virtual image $9\frac{1}{2}$ inches from the lens, the greatest amount of magnification will be obtained, viz.—

$$M = \frac{l}{Kb}\left(1 + \frac{q}{f'}\right) = 1 + \frac{9\frac{1}{2}}{4} = 3\frac{3}{8}$$

There is another way in which a convex lens may be used as a magnifying glass, viz. when the object is placed in the first focal plane of the lens (Fig. 35). HOH′ is the

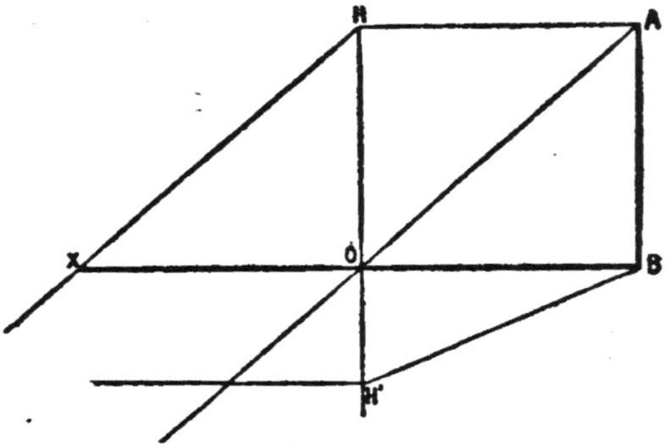

FIG. 35.

principal plane of the lens, which it is unnecessary to indicate, and the object BA is placed in the first focal plane, so that the incident cone of rays from B (H′BO) will, after refraction, emerge as a beam parallel to the axis, while the incident cone from A (OAH) will emerge as a beam parallel to AO. If, now, an emmetropic eye (or an eye adapted to see very distant objects) be situated behind the lens, an image of BA will be formed on its retina. It is clear that in this case the apparent size of BA will be independent of the position of the eye, for its size depends on tan BOA or tan θ''. If, however, the eye be situated behind the point X, it is obvious that it could not see the whole of the object, as only rays from the central part of BA would enter the eye; in fact, on increasing the distance of the eye from the lens the *field of view* would be diminished, but there would be no alteration of the magnification of that part that was seen.

As BA is in the first focal plane, $\tan \theta'' = \frac{o}{f'}$, and so the magnification M′ in this case is $\frac{\tan \theta''}{\tan \theta} = \frac{o}{f'} \cdot \frac{l}{o} = \frac{l}{f'}$.

Our lens of 4 inches focus, used in this way, would only give a magnification of $\frac{l}{f'}$, or $\frac{10}{4} = 2{\cdot}5$, but it would be less fatiguing to the eye.

On comparing the magnifying power of a convex lens used in these two different ways, we see that in the first case it is $1 + \frac{q}{f'}$, and in the second case it is $\frac{l}{f'}$, or $\frac{d+q}{f'}$. Therefore, if d be less than f' the first method gives the higher magnification, and *vice versâ*.

Graphic Method for Lenses.—A method similar to that illustrated in Figs. 28 and 29 may be employed for finding the position and the relative size of the image formed by any lens. Fig. 28, when F″F′ is made equal to F′H, will represent the construction necessary for a convex lens whose first focal distance (f') is indicated by F′H, whereas Fig. 29 if inverted would give the construction for a concave lens when F″F′ is made equal to F′H, its focal distance. Such methods will, however, be seldom found to be of practical use, as we have such simple formulæ as $q = \frac{pf''}{p - f'}$, and $\frac{i}{o} = \frac{f'}{f' - p}$ ready to our hand for the solution of questions like these.

Power of a Lens. Dioptres.—In a previous section we used the expression "magnifying power," and we saw that it varied inversely as f'; we have seen, too, that strong lenses with strongly curved surfaces had short focal lengths. We are therefore quite ready to admit that $\frac{1}{f'}$ is an adequate measure of the power of a lens. The unit universally adopted is that of a lens of one metre focal length, which is called a dioptre, and is denoted by the symbol D. We know that a convex lens of 25 cm. focal length is four times stronger than one of 100 cm. focal length. This is very simply expressed by calling the former lens +4D and the latter

$+1D$. Note that it is the first focal length, f', that is considered and that determines the sign of the lens. We give an example or two, so as to make this nomenclature quite clear.

What is the power of the following lenses in dioptres?—two concave lenses of 10 cm. and 20 cm. focal length respectively, and one convex lens of 80 cm. and another convex lens of 22·5 inches focal length.

The first focal distance of a concave lens is negative, so in the first case $f' = -10$ cm., or $-\frac{1}{10}$ metre; $\frac{1}{f'}$ is then $-10D$; in the second case $f' = -20$ cm. $= -\frac{1}{5}$ metre, so $\frac{1}{f'}$ is $-5D$.

The first focal distance of a convex lens is positive, so if $f' = 80$ cm., the power of the lens in dioptres is $\frac{100}{80}$, or $+1{\cdot}25D$.

In every case the dioptric power is given by $\frac{1}{f'}$, in metres, or $\frac{100}{f'}$ in centimetres. When we have f' given in inches, we must convert it into metres. As 39·37... inches are equivalent to one metre, 22·5 inches are equivalent to $\frac{22{\cdot}5}{39{\cdot}37}$ metres, so $\frac{1}{f'}$ in metres is $\frac{39{\cdot}37}{22{\cdot}5} = 1{\cdot}75$. The dioptric power of this last glass is then $+1{\cdot}75D$.

Thin Lenses in Juxtaposition.—A succession of thin lenses, such as that shown in Fig. 36, has practically the same effect as that of one lens whose power is represented by the sum of the dioptric strengths of the components of the system. Let us suppose that the first lens (a meniscus) is $+1D$, the second lens $+10D$, and the third lens $-9{\cdot}5D$; the sum of the dioptric strengths is $+1 + 10 - 9{\cdot}5$, or $+1{\cdot}5D$. Consequently, the system will be practically equivalent to one lens of $+1{\cdot}5D$. This illustrates the extreme simplicity that results from denoting lenses by their power instead of by their focal length. If the thickness of the lenses, or of their combination, had to be taken into account, a correction would have to be added, which will be explained when dealing with cardinal points.

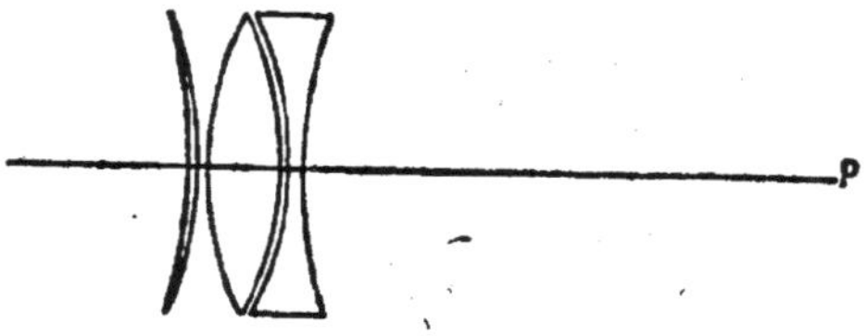

FIG. 36.

When a space separates the several lenses of the system, the rule just stated does not apply; for such a case the reader is referred to p. 91.

Optical Centre.—We have frequently made use of the term "the centre of a lens," and the reader most probably thinks that it is equivalent to the mid-point of the thickness of the lens. It is so in symmetrical biconvex or biconcave lenses, but it may be outside the lens altogether, as O in Fig. 37. The optical centre of a lens may be defined as that point in which the line joining the extremities of any parallel radii of the two bounding surfaces cuts the axis.

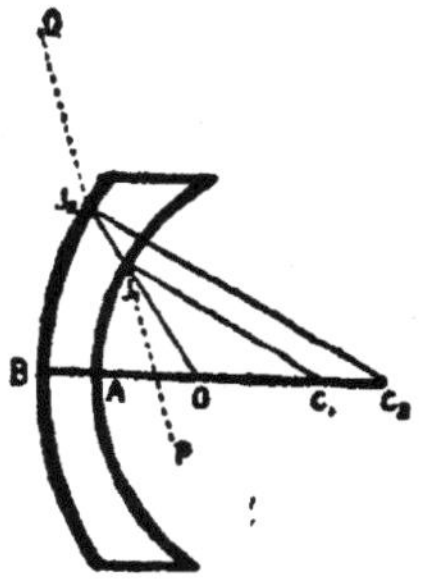

FIG. 37.

In Fig. 37, BAC_1C_2 is the axis, and C_1J_1 and C_2J_2 are two parallel radii. The line joining J_2 and J_1, if produced, cuts the axis in O, which is then by the definition the optical centre.

The point O is a fixed point, the position of which depends only on the lengths of the radii and the thickness t of the lens; for by similar triangles

$$\frac{C_1O}{C_2O} = \frac{C_1J_1}{C_2J_2} = \frac{r_1}{r_2}$$

Also
$$\frac{r_1}{r_2} = \frac{C_1O}{C_2O} = \frac{r_1 - OA}{r_2 - OB} = \frac{OA}{OB}$$

$$\therefore \frac{OB - OA}{OB} \text{ or } \frac{AB}{OB} = \frac{r_2 - r_1}{r_2}$$

So $$OB = \frac{r_2 t}{r_2 - r_1} \quad \text{and} \quad OA = \frac{r_1 t}{r_2 - r_1} \quad . \quad (a)$$

Consequently, in biconvex and biconcave lenses, when either r_1 or r_2 is negative, $r_2 - r_1 > r_2$, so $AB > OB$; in other words, O is within the lens. The optical centre O has the following important property: any incident ray, such as PJ_1, passing through the lens so that its direction while within the lens passes (either actually or virtually) through the centre O will on emerging from the lens, have a direction J_2Q parallel to its direction PJ_1 when incident; and, conversely, if any emergent ray be parallel to its corresponding ray, it must, while within the lens, have assumed a direction that would pass through the optical centre. This property of the optical centre follows at once from the fact that the tangents at the two points where refraction takes place are parallel, and therefore the effect on this ray is the same as that due to refraction through a plate (p. 40), *i.e.* that the angle of emergence is always equal to the angle of incidence ϕ. Pencils that pass either actually or virtually through the optical centre are called *centric pencils.*

It is an easy matter to find the optical centre of any lens from the expression (*a*), if we are given the thickness *t*, (AB), and the radii of the two surfaces. It is quite immaterial which way the light is supposed to be travelling, whether from P to Q or from Q to P (Fig. 38). In the case of a biconvex or a biconcave lens, the ray while within the lens actually (not virtually) does pass through O.

Let us consider the left surface as A, the first surface, then if $r_1 = 3$, $r_2 = -4$, and $t = 2$, by (*a*) we know that

$$OA = \frac{r_1 t}{r_2 - r_1} = \frac{3 \times 2}{-4 - 3} = -\frac{6}{7}$$

and $$OB = \frac{r_2 t}{r_2 - r_1} = \frac{-4 \times 2}{-4 - 3} = \frac{8}{7}$$

and we notice that O is situated within the lens and nearer the most strongly curved surface. Now, an incident ray PK′ that emerges in a parallel direction, as K″Q, will not follow the course indicated in Fig. 38 within the lens, for it will be bent at the first surface, then pursue a straight course through O, and then be again bent on emergence in the direction K″Q. It is, however, very convenient for purposes of calculation to find two points, K′ and K″, on the axis which will save us the trouble of calculating the angle of obliquity of the path within the lens. This can clearly be done by continuing the incident and emergent rays until they intersect the axis in K′ and K″, or the nodal points. This is a simple geometrical method, but it does not readily lead to an analytical expression, and the accuracy of the result obtained by a geometrical method depends, of course, upon the accuracy and size of the drawing. There are, however, two analytical methods for locating the nodal points

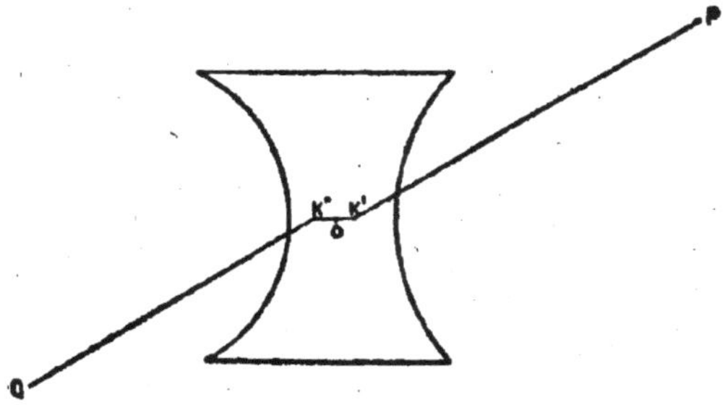

FIG. 38.

that we shall presently give, but we shall first deal with a case in which the nodal points coincide at the optical centre.

* **Refraction of a Sphere. Coddington Lens.**—The sphere may be considered as a kind of double convex lens, and there are certain advantages attending its use, which we shall investigate.

Note, first, that all pencils (even those that are oblique) which traverse the centre of the sphere pass normally into it, so that in this case all centric pencils are also axial.

It may be readily seen, from (*a*), p. 83, that the geometrical centre of the sphere is also the optical centre of the lens; so it will be convenient to find an expression for the focal distances when considered as measured towards the centre.

Let O (Fig. 39) denote the centre of the sphere, and let P denote an object in front of it, Q_1 its image due to refraction at the surface A, and Q the final image due to the refraction of Q_1 at the second surface, B.

Let P denote PO = PA − OA = $p - r_1$,
and q' „ $Q_1O = Q_1A - OA = q_1 - r_1$,
and let Q „ QO = QB − OB = $q - r_2$.

By the formula of p. 53,

$$\frac{p-r_1}{q_1-r_1}=\frac{\mu'}{\mu_0}\cdot\frac{p}{q_1} \quad \text{or} \quad \mu_0 q_1(p-r_1)=\mu' p(q_1-r_1)$$

or

$$\mu_0(q'+r_1)P=\mu'(P+r_1)q'$$

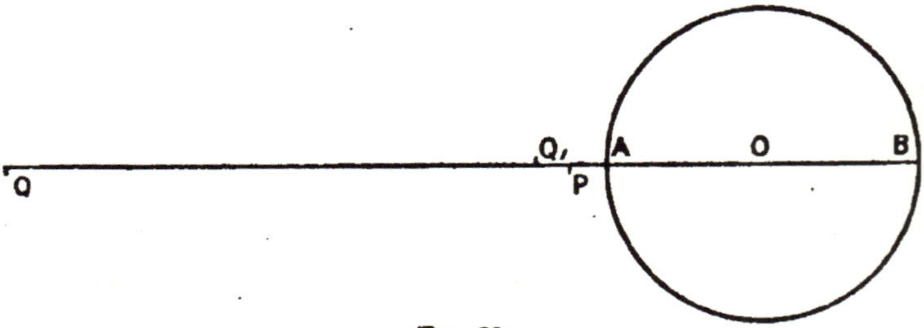

Fig. 39.

Dividing by $\mu' P q' r_1$ we obtain

$$\frac{\mu_0}{\mu' r_1}+\frac{\mu_0}{\mu' q'}=\frac{1}{r_1}+\frac{1}{P} \quad . \quad . \quad . \quad . \quad . \quad (a)$$

Now regarding Q_1 as the object for refraction at the second surface B, we note that p is to be replaced by $Q_1B = Q_1O + OB = q' + r_2$, and $\frac{\mu'}{\mu_0}$ by $\frac{\mu_0}{\mu'}$ in the formula $\frac{p-r}{q-r}=\frac{\mu'}{\mu_0}\cdot\frac{p}{q}$.

So $$\frac{q'}{q-r_2}=\frac{\mu_0}{\mu'}\cdot\frac{q'+r_2}{q} \quad \text{or} \quad \frac{q'}{Q}=\frac{\mu_0}{\mu'}\cdot\frac{q'+r_2}{Q+r_2}$$

or

$$\mu_0 Q(q'+r_2)=\mu' q'(Q+r_2)$$

or dividing by $\mu' Q q' r_2$—

$$\frac{\mu_0}{\mu' r_2}+\frac{\mu_0}{\mu' q'}=\frac{1}{r_2}+\frac{1}{Q}$$

On subtracting (*a*)—

$$\frac{\mu_0}{\mu' r_1} + \frac{\mu_0}{\mu' q'} = \frac{1}{r_1} + \frac{1}{P}$$

we have

$$\frac{\mu_0}{\mu'}\left(\frac{1}{r_2} - \frac{1}{r_1}\right) = \frac{1}{r_2} - \frac{1}{r_1} + \frac{1}{Q} - \frac{1}{P}$$

or

$$\frac{1}{Q} - \frac{1}{P} = \frac{\mu_0 - \mu'}{\mu'}\left(\frac{1}{r_2} - \frac{1}{r_1}\right) \quad . \quad (b)$$

Noting that $r_2 = -r_1$, we see that $\frac{1}{r_2} - \frac{1}{r_1} = \frac{2}{r_2}$, and by making P infinite we can find f''; similarly by making Q infinite we can find f'. We have then

$$\frac{1}{f''} = \frac{\mu_0 - \mu'}{\mu'} \cdot \frac{2}{r_2} \quad \text{and} \quad \frac{1}{f'} = \frac{\mu' - \mu_0}{\mu'} \cdot \frac{2}{r_2}$$

and so we may write (*b*) as

$$\frac{1}{Q} - \frac{1}{P} = \frac{1}{f''} \quad \text{or} \quad \frac{f'}{P} + \frac{f''}{Q} = 1$$

This tedious work might have been avoided had we been able to use the method of "Cardinal Points," but it is well to see the labour that is involved even in the simple case of a sphere, if we are ignorant of the better method that we shall shortly describe. It has been taken as an instance in which our old formula $\frac{f'}{p} + \frac{f''}{q} = 1$ can be used if all the distances involved are measured towards a certain point, in this case the centre of the sphere. The section on Cardinal Points will show that in any system, however complex, two points can be found such that, if the appropriate distances are measured towards them, the formula $\frac{f'}{p} + \frac{f''}{q} = 1$ will hold good. In the case of a sphere these two points are coincident.

The Coddington lens, represented in Fig. 40, is a very convenient form of pocket magnifying glass, and has this important advantage over an ordinary convex lens—the peripheral parts of the virtual image are as distinct as the central parts, provided that there is a central stop, so that none but centric rays traverse it. Usually a deep equatorial groove is ground in the lens, as indicated by the shaded part in the figure. In practice it is found that the central aperture must not be greater than a fifth of the focal length of the lens. The defects of the lens are: (1) the image is curved as the peripheral parts of the object are further away from O than the central parts; (2) the working distance is very short; and (3) the field of view is limited; as only those emergent pencils which can enter the pupil of the eye are effective; it is advisable, therefore, to bring the lens as close to the eye as possible.

The Stanhope lens is somewhat similar to the Coddington; it is a short glass cylinder with its ends ground convex to an

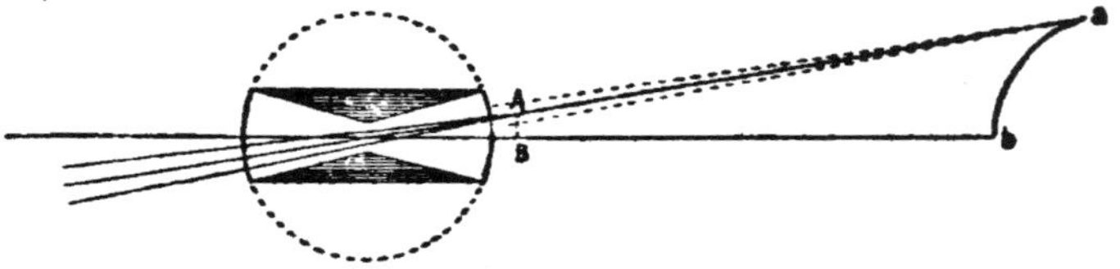

FIG. 40.

unequal degree of curvature. The object is placed on the surface of lesser curvature, and the length of the cylinder is such that when the more convex surface is turned towards the eye, a distinct magnified image of the object is seen.

Nodal Points of a Thick Lens.—The following method is applicable only to lenses, and can, further, only be used when the initial and final media are the same. Should they be different, as in the case of the eye, it will be necessary to use the method described later (p. 91), which is perfectly general. The only advantage of this method is its suitability in the case of thick lenses, if we forget the formulæ for Cardinal Points.

Given the thick lens, we first find the position of O by (*a*), p. 83. We then find the position of the image of O as viewed from the incident surface (A), and call it K′, the first nodal point. We then find the position of the image of O when viewed from the other surface (B), and call it K″, the second nodal point. A warning must be given when dealing with menisci, as the point O must always be considered as situated in the glass, even when it is not, as in Fig. 37, or totally erroneous results will be obtained. Consequently the method must be regarded simply as a trick or device by which it can be mathematically proved that accurate results will be obtained.

We will take the case of Fig. 38 as an example, where $r_1 = 3$, $r_2 = -4$, $t = 2$, and OA and OB we have found to be $-\frac{6}{7}$ and $\frac{8}{7}$ respectively. We are now considering the left-hand surface to be the incident surface A, so that light is supposed to be travelling from Q to P. The problem is the same as that on p. 57; we first find the image of O formed by the surface A. As O is in the glass, $\mu_0 = 1{\cdot}5$ and $\mu' = 1$, so—

$$f_1' = \frac{-\mu_0 r_1}{\mu' - \mu_0} = \frac{-1{\cdot}5 \times 3}{1 - 1{\cdot}5} = 9, \text{ while } f_1'' = \frac{-\mu'}{\mu_0} f_1' = \frac{-9}{1{\cdot}5} = -6$$

$$\therefore \text{K}'\text{A or } q = \frac{pf_1''}{p - f_1'} = \frac{-\frac{6}{7}(-6)}{-\frac{6}{7} - 9} = \frac{36}{-69} = -\frac{12}{23}$$

Similarly for the surface B—

$$f_2' = \frac{-\mu_0 r_2}{\mu' - \mu_0} = \frac{-1{\cdot}5(-4)}{1 - 1{\cdot}5} = \frac{6}{-0{\cdot}5} = -12,$$

and $$f_2'' = -\frac{\mu'}{\mu_0} f_2' = \frac{12}{1{\cdot}5} = 8$$

And as OB or $p' = \frac{8}{7}$

$$\text{K}''\text{B or } q' = \frac{p'f_2''}{p' - f_2'} = \frac{\frac{8}{7} \times 8}{\frac{8}{7} + 12} = \frac{64}{92} = \frac{16}{23}$$

The position of the nodal points is quite independent of the direction in which the light is travelling; it is only their names (first or second) that are changed. We have been considering light passing from Q to P, so K′ refers to the left-hand surface. In the diagram the reverse condition is indicated; the light is passing from P to Q, so that the right-hand surface is the incident surface, and the nodal point corresponding to it is called K′, while K″ refers to the emergent (left) surface.

The mode of using these nodal points will be illustrated later, when we deal with the geometrical construction of the image of a complex system.

Spherical Aberration.—In Fig. 41 a beam of parallel rays is shown that encounters a double prism; the more central rays SI and S′I′ intersect at R after traversing the prism,

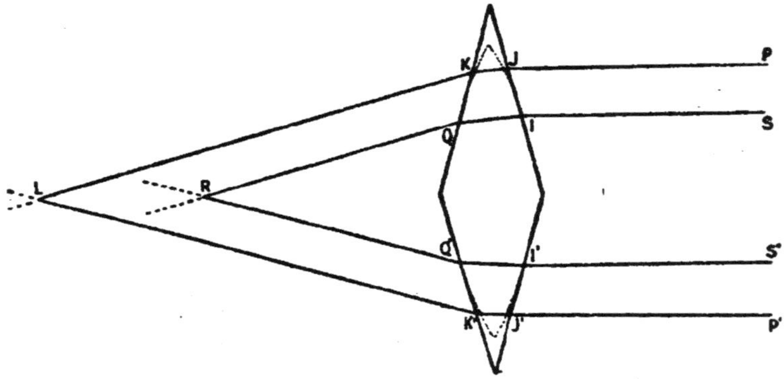

Fig. 41.

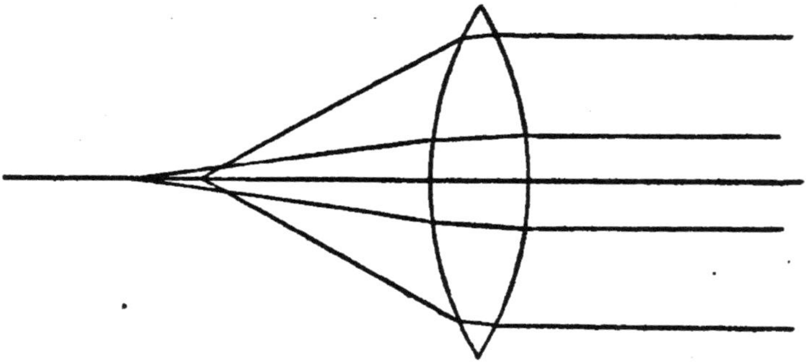

Fig. 42.

while the more peripheral rays PJ and P′J′ intersect at a more distant point, L. Clearly the aberration RL might be, obviated by bevelling the peripheral parts of the prism, so that the incident rays at J and J′ would undergo a greater deviation. In a spherical lens (Fig. 42) this bevelling is carried out so that it acts like a prism, the strength of which is continually increasing from the axis to the periphery. Unfortunately, a spherical surface is not quite the right shape—indeed, the bevelling has been carried too far, for the peripheral parts of an incident parallel beam intersect at a closer point than the more central parts. The diagram illustrates what is called spherical aberration (undercorrected), while Fig. 41 will indicate what is called overcorrection of spherical aberration.

There is one obvious way of making the focus of such a lens more definite, *i.e.* by cutting off the aberrant peripheral rays with a stop, so that the focus is only formed by the intersection of the thin axial pencil. Another method which has certain advantages, as has been pointed out by Lord Rayleigh, is to block out the central rays and use only the peripheral ones. This, however, has been rarely used in practice.

Just as with reflection at a mirror, when an oblique or eccentric incident pencil is considered, the refracted pencil is astigmatic, and presents the same focal lines with the same sphenoid shape between them. All this has been omitted in the diagram for simplicity. The short line on the axis between the focus of the peripheral and that of the more central rays may be regarded as the secondary focal line, while the intersection of the peripheral with the central rays indicates the position of the primary focal line sagittal to the plane of the paper.

It may reasonably be asked, Why are lenses made of this erroneous shape? The answer is that it is impossible to mould glass of the right shape with any approach to accuracy, and grinding by hand to any shape but spherical is a most laborious and difficult undertaking.

If we confine our attention to thin oblique pencils, we see that they may be of two kinds: (1) an oblique eccentric pencil that is incident upon a peripheral or eccentric part of the lens; and (2) an oblique centric pencil that traverses (either actually or virtually) the optical centre of the lens, as PQ in Fig. 38 or PQ in Fig. 37. The exact mathematical investigation of the form of pencil after (1) oblique eccentric refraction through a lens is most tedious and difficult, and does not lead to any simple approximate expression; and, further, it is of little practical importance, except to certain instrument makers, so we must refer the scientific mathematician to more advanced treatises on this point. (2) Oblique centric refraction will be briefly treated in the Appendix (p. 121), as it is of considerable practical importance.

***Cardinal Points.**—Gauss has shown us how to extend the use of the formulæ $\frac{f'}{p}+\frac{f''}{q}=1$, etc., not only to thick lenses, but also to any refracting system however complex, formed of any number of media bounded by centred spherical surfaces. The only requisite is to find the position of two points called the Principal Points of the system under consideration. When the distance of the First Principal Point H′ from the object P is denoted by PH′, and the distance of the Second Principal Point H″ from the object Q is denoted by QH″, we have $\frac{F'H'}{PH'}+\frac{F''H''}{QH''}=1$ universally true, where F′H′ is the distance of H′ from the First Principal Focus F′, and where F″H″ is the distance of H″ from the Second Principal Focus F″ (Fig. 44).

Before actually dealing with the problem, we will show exactly what it is that we want to find. Let H_1A and H_2B represent the principal planes of two thin concave lenses (Fig. 43). An incident ray parallel to the axis will, on traversing the concave lens at H_1, be refracted in the direction F_1H_2 as if it were proceeding from the second focus of the first lens F_1; on now meeting the second lens at H_2, it will again be deviated in the direction H_2R as if it were proceeding from N. We wish to find a lens which shall have an equivalent action to these two lenses. It is clear from the diagram that a concave lens placed in the position KX, if its second focus be at N_1, will have precisely the same effect on the incident ray SH_1K as the combination of lenses had on SH_1. The whole object of Gauss's method is to find the position of X, *i.e.* the situation of the equivalent lens, and also its focal distance NX. In such a simple case as this, it is an easy problem to solve by purely geometrical means, but Gauss has shown us how to deal with any system, however complex.

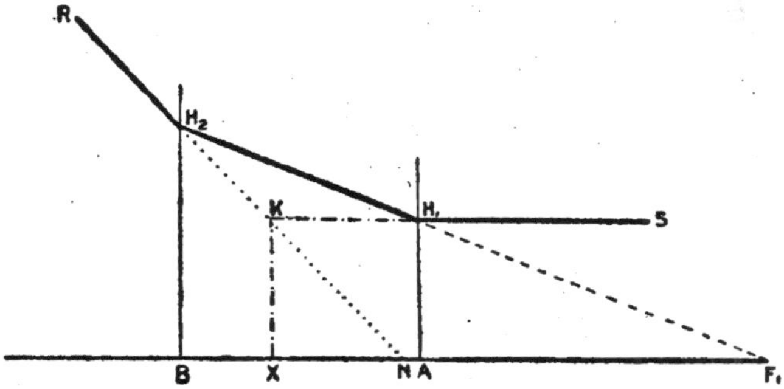

FIG. 43.

The method used by Gauss is intricate and involves a considerable knowledge of mathematics, but as we know the result of his calculations, we shall be able to find the positions of H′ and H″ by easy algebra if we treat the question in the right way. In any algebraic problem it is necessary to pay the utmost attention to the algebraic statement of the problem, but after it is once correctly stated, think no more about the meaning of the future operations until you get your result. "Put it into the algebraic mill and turn the handle." The chief difficulty of all beginners in mathematics is that they try and think what each algebraic step means. This is quite unnecessary; thought is only required when stating the problem.

We will take as an example the thick lens in Fig. 44, and use the followiug symbols to denote the various distances.

The thickness of the lens AB $= t$, the distances of A from the first (f_1') and second (f_1'') foci, due to the first refraction at A are $f_1'A, = f_1'$ and $f_1''A = f_1''$, and similarly when f_2' and f_2'' denote the first and the second foci due to the second refraction at B, $f_2'B = f_2'$, and $f_2''B = f_2''$. The values of these symbols can be determined in any given case by the well-known formulæ of p. 55; they are indicated in the special case illustrated in the diagram by the letters below the line, whereas H′ and H″ that we have to find are shown above the line. We shall also use the symbols h' and h'' to denote the distances H′A and H″B respectively.

Now, if P denote any object (not illustrated in the diagram), PA or p will denote the distance of the surface A

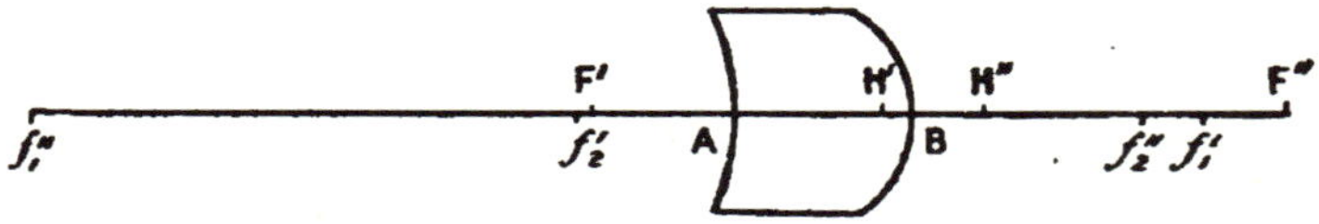

FIG. 44.

from it, and we can easily find the position of the image Q_1, due to the first refraction, for if q_1 denote the distance Q_1A,

$$q_1 = \frac{pf_1''}{p - f_1'}$$

Regarding now Q_1 as the object that forms the final image Q by refraction at the second surface B, we can as easily express Q_1B in terms of QB or q. For

$$Q_1B = \frac{qf_2'}{q - f_2''}$$

But $Q_1B = Q_1A + AB = q_1 + t$, so we can eliminate q_1 by substituting for it our previous expression. We have then

$$\frac{pf_1''}{p - f_1'} + t = \frac{qf_2'}{q - f_2''} \quad . \quad . \quad . \quad . \quad (a)$$

We have to find the positions of H′ and H″ so that the formula

$$\frac{F'H'}{PH'} + \frac{F''H''}{QH''} = 1$$

shall give a result that is identical with that given by (a).

Now, $PH' = PA + AH'$, or $PA - H'A$, *i.e.* $p - h'$; similarly, $QH'' = QB - H''B = q - h''$, and if we denote

F'H' by F' and F''H'' by F'', this last formula can be written—

$$\frac{F'}{p-h'}+\frac{F''}{q-h''}=1 \quad . \quad . \quad . \quad . \quad . \quad (b)$$

All, then, that we have to do is to make these two formulæ (a) and (b) identical with each other. For (a) we shall write—

$$pqf_1'' - pf_1''f_2'' + t(pq - pf_2'' - qf_1' + f_1'f_2'') = pqf_2' - qf_1'f_2'$$

or

$$pq(t+f_1''-f_2')-p(f_1''f_2''+tf_2'')+q(f_1'f_2'-tf_1')+tf_1'f_2''=0 \quad (a')$$

And for (b) we shall write

$$F'q - F'h'' + F''p - F''h' = pq - ph'' - qh' + h'h''$$

or

$$pq - p(h'' + F'') - q(h' + F') + h'h'' + F'h'' + F''h' = 0 \;. \quad (b')$$

On comparing the coefficients of (a') and (b') we see that the two expressions will be identical if

$$t+f_1''-f_2' = \frac{f_1''f_2''+tf_2''}{h''+F''} = \frac{tf_1'-f_1'f_2'}{h'+F'} = \frac{tf_1'f_2''}{h'h''+F'h''+F''h'}$$

or calling this expression K,

$$\text{if } h''+F'' = \frac{f_1''f_2''+tf_2''}{K}, \quad \text{if } h'+F' = \frac{tf_1'-f_1'f_2'}{K}$$

$$\text{and if } h'h''+F'h''+F''h' = \frac{tf_1'f_2''}{K}$$

$$i.e. \text{ if } h' = \frac{f_1't}{K}, \text{ if } h'' = \frac{f_2''t}{K}, \text{ if } F' = \frac{-f_1'f_2'}{K}, \text{ and if } F'' = \frac{f_1''f_2''}{K}$$

these values are clearly consistent, and they are therefore the solutions required.

In the case illustrated in the diagram (Fig. 44), the radius of the first surface, $r_1 = 4$, that of the second surface $r_2 = 2$, $\mu_0 = 1$, $\mu' = 1{\cdot}5$, and $t = 3$.

$$\therefore f_1' \text{ or } \frac{-\mu_0 r_1}{\mu' - \mu_0} = \frac{-4}{1{\cdot}5 - 1} = -8\,;$$

$$f_1'' \text{ or } \frac{-\mu'}{\mu_0} f_1' = (-1{\cdot}5)(-8) = 12\,;$$

$$f_2' \text{ or } \frac{-\mu' r_2}{\mu_0 - \mu'} = \frac{-(1{\cdot}5)(2)}{1 - 1{\cdot}5} = 6\,;$$

$$f_2'' \text{ or } \frac{-\mu_0}{\mu'} f_2' = \frac{-6}{1{\cdot}5} = -4\,;$$

$$\text{and } K = t + f_1'' - f_2' = 3 + 12 - 6 = 9$$

$$\therefore h' = \frac{f_1' t}{K} = \frac{-8 \times 3}{9} = -2\tfrac{2}{3}\,;$$

$$h'' = \frac{f_2'' t}{K} = \frac{-4 \times 3}{9} = -1\tfrac{1}{3}$$

$$F' = -\frac{f_1' f_2'}{K} = -\frac{-8 \times 6}{9} = 5\tfrac{1}{3}\,;$$

$$F'' = \frac{f_1'' f_2''}{K} = \frac{12(-4)}{9} = -5\tfrac{1}{3}$$

If in this case we were to use the formula on p. 70, which neglects the thickness of the lens, we should have—

$$\frac{1}{f''} = \overline{\mu - 1}\left(\frac{1}{r_1} - \frac{1}{r_2}\right) = \frac{1}{2}\left(\frac{1}{4} - \frac{1}{2}\right) = -\frac{1}{8}$$

or $f'' = -8$, and f' or $-f'' = 8$, a result that is quite erroneous. However, on adding the correcting term $-\frac{(\mu-1)^2 t}{\mu r_1 r_2}$, or on using the form—

$$\frac{1}{f''} = \frac{\mu - 1}{r_1 r_2}\left(r_2 - r_1 - \frac{\overline{\mu - 1}\,t}{\mu}\right), \text{ we get } \frac{1}{f''} = \frac{1}{16}\left(2 - 4 - \frac{3}{3}\right) = -\frac{3}{16}$$

$$\therefore f'' \text{ or } -f' = -5\tfrac{1}{3}, \text{ which is correct.}$$

When two compound systems, A and B, are combined, A being the first system traversed by the incident light, $t = H_a''H_b'$, *i.e.* the distance of the first principal point of the second system (B) from the second principal point of the first system (A); whereas $h' = H'H_a'$, or the distance of the first principal point of the complete system measured towards the first principal point of A. Similarly $h'' = \mathrm{H}_a''\mathrm{H}_b''$ and $K = \mathrm{H}_a''\mathrm{H}_b' + f_a'' - f_b'$, the subscripts a and b denoting the systems to which the letters refer.

These formulæ are a little difficult to remember, but when reference can be made to them they are easy to employ, and they are of universal application. We shall give three examples of their use, to illustrate their simplicity and value.

Ex. (1).—The ordinary form of eyepiece for a microscope (Huygenian) consists of two convex lenses, the distal one being called the field-glass, and the proximal one, to which the eye is applied, the eyeglass. It is found that a No. III. eyepiece, with an alleged magnification of eight diameters, cannot be used as a magnifying glass when held in the normal position before a microscope slide. Explain this, and show how it can be used as a magnifying glass.

In all Huygenian eyepieces the field-glass (A) has a focal length three times that of the eyeglass (B), and the distance between them is one-half of the sum of the focal distances.

The advantages obtained by this construction cannot be explained in this place, but it may be stated that the spherical aberration of the system is less when each lens has its second focus at the same point, and that the size (though not the position) of the image is achromatized when $t = \frac{1}{2}(f_a' + f_b')$.

With a No. III. eyepiece $f_a' = \frac{5}{2}$ ins., and $f_b' = \frac{5}{6}$ in. As we may regard the lenses of negligible thickness, we have $t = \frac{5}{3}$ ins.,

and K or $t + f_a'' - f_b' = \frac{5}{3} - \frac{5}{2} - \frac{5}{6} = -\frac{5}{3}$

$$\therefore h' = \frac{f_a't}{K} = \frac{\frac{5}{2} \times \frac{5}{3}}{-\frac{5}{3}} = -\frac{5}{2}$$

$$h'' = \frac{f_b''t}{K} = \frac{-\frac{5}{6} \times \frac{5}{3}}{-\frac{5}{3}} = \frac{5}{6}$$

$$F' = -\frac{f_a'f_b'}{K} = -\frac{\frac{5}{2} \times \frac{5}{6}}{-\frac{5}{3}} = \frac{5}{4}, \quad \text{and } F'' = -\frac{5}{4}$$

The position of the cardinal points of this Huygenian eyepiece is illustrated in Fig. 45, and it will be seen that the first principal focus (F′) is within the eyepiece; and we know from p. 78 that the object must be placed not further off than F′, so clearly the eyepiece will not act as a magnifying glass when used in this way. However, F″ is situated outside the lens, beyond B, so that if we reverse the eyepiece, using A as the eyeglass and putting the microscope slide in the position of F″, it will form a very efficient magnifying glass. In such a position, when the eyepiece is reversed, F″ and F′ are simply interchanged. We have shown that the magnifying power of a lens is measured by $\frac{l}{f'}$. In this case $\frac{l}{f'} = 10 \div \frac{10}{8} = 8$.

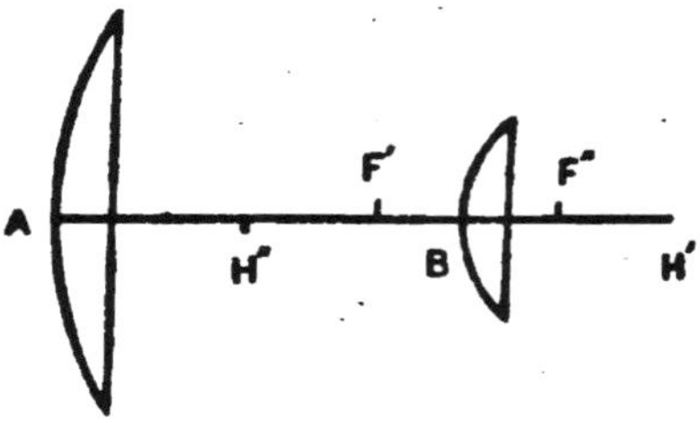

Fig. 45.

When a Huygenian eyepiece is used in a microscope, the image formed by the objective of the instrument is formed within the eyepiece at F′, so that magnification results normally.

If the two lenses of the eyepiece were placed in contact, the power of the combination would be

$$\frac{1}{f_a'} + \frac{1}{f_b'} = \frac{2}{5} + \frac{6}{5} = \frac{8}{5}$$

or the combination would then be equivalent to a lens of $\frac{5}{8}$ in. focal length, while its magnifying power would be doubled, being 16; it would, however, manifest all the chromatic and aberrational errors that the Huygenian eyepiece in part corrects.

Ex. (2).—Suppose that the eyeglass in a No. III. Huygenian eyepiece were replaced by a concave lens of equal strength: what would be the power of the combination, and what use could be made of the instrument?

Here $f_b' = -\frac{5}{6}$ and K or $t + f_a'' - f_b' = \frac{5}{3} - \frac{5}{2} + \frac{5}{6} = 0$

and as $F' = -\frac{f_a' f_b'}{K} = \infty$, the power $\frac{1}{F'} = \frac{1}{\infty}$.

The power of the combination is therefore 0, *i.e.* all incident parallel light emerges parallel, and it might be thought to be useless as an instrument; but a little consideration will show that it will act as a very efficient telescope—indeed, this is the form of the Galilean telescope or opera glass.

A very distant object, subtending a visual angle θ, will also make the angle $f_a''AD$ at the field-glass (Fig. 46) equal

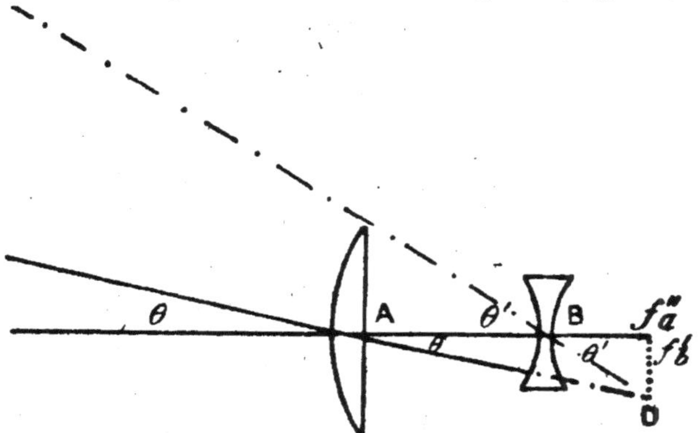

FIG. 46.

to θ, and an image of the object would be formed in the second focal plane at $f_a''D$. But $f_a''D$ is also the situation of the first focal plane of B, consequently the incident parallel light which is converging towards D will, owing to the interposition of the concave lens at B, proceed as a beam of parallel rays in the direction BD (the spaced and dotted lines). Similarly, the axial incident rays which, after traversing A, tend to converge towards f_a'' will, owing to the action of B, proceed as a parallel beam in the direction of the axis. Note, however, that the direction of the oblique pencil has been changed, as it now makes an angle θ' with the axis instead of θ. An eye, therefore, if adjusted for parallel rays, placed close to B will see an erect virtual image subtending the angle θ', and the magnification of the instrument will be—

$$\frac{\tan\theta'}{\tan\theta} = \frac{f_a''A}{f_b'B} = \frac{-5}{2} \div \frac{-5}{6} = 3$$

Ex. (3).—We will now find the cardinal points of the human eye. According to Tscherning's most recent investigations, the lens of the eye has a focal length (ϕ) of 51·34 mm., and its principal points, indicated as H_1 and H_2 in Fig. 47, are so situated that $H_1A_1 = -2{\cdot}308$ mm., and $H_2A_2 = 1{\cdot}385$ mm. The thickness of the lens is 3·9 mm., and it is placed 3·6 mm. behind the cornea (A_0). We shall consider the media to be of the same refractive index 1·3375, bounded by the cornea of radius $-7{\cdot}8$ mm. In the diagram

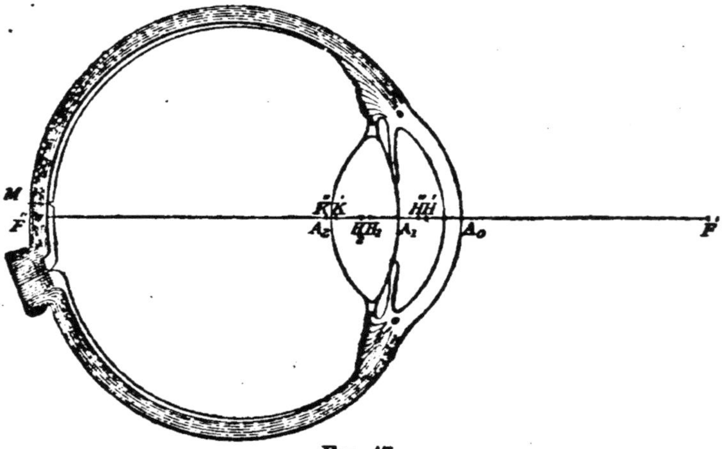

FIG. 47.

the incident light is presumed to be travelling from right to left, so we will regard this as the positive direction.

Taking first the corneal refraction, we find

$$f' = \frac{-r}{\mu - 1} = \frac{+7{\cdot}8}{{\cdot}3375} = 23{\cdot}11 \text{ mm.}$$

and $$f'' = -\mu f' = -30{\cdot}91 \text{ mm.}$$

The distance t or

$$A_0H_1 = A_0A_1 + A_1H_1 = 3{\cdot}6 + 2{\cdot}308 = 5{\cdot}908 \text{ mm.}$$

and K or $t + f'' - \phi' = 5{\cdot}908 - 30{\cdot}91 - 51{\cdot}34 = -76{\cdot}342$

$$\therefore H'A_0, \textit{ i.e. } h' \text{ or } \frac{f't}{K} = \frac{23{\cdot}11 \times 5{\cdot}908}{-76{\cdot}342} = -1{\cdot}789 \text{ mm.}$$

$$H''H_2, \textit{ i.e. } h'' \text{ or } \frac{\phi''t}{K} = \frac{-51{\cdot}34 \times 5{\cdot}908}{-76{\cdot}342} = 3{\cdot}973 \text{ mm.}$$

$$F'H' \text{ or } F' = \frac{-f'\phi'}{K} = \frac{-23{\cdot}11 \times 51{\cdot}34}{-76{\cdot}342} = 15{\cdot}54 \text{ mm.}$$

$$F''H'' \text{ or } F'' = \frac{f''\phi''}{K} = \frac{-30{\cdot}91(-51{\cdot}34)}{-76{\cdot}342} = -20{\cdot}79 \text{ mm.}$$

The distance of the second principal point from the cornea can be easily obtained, for

$$H''A_0 = H''H_2 + H_2A_2 + A_2A_1 + A_1A_0,$$

or $$H''A_0 = 3{\cdot}973 + 1{\cdot}385 - 3{\cdot}9 - 3{\cdot}6 = -2{\cdot}142 \text{ mm.}$$

The point H″ is therefore only 0·353 mm. from H′. We shall not, then, introduce any appreciable error in considering that they coincide in one point H, towards which both the focal distances are measured.

Nodal Points of a Complex System.—When we have found the cardinal points by the above method, it is a simple matter to find the nodal points. Take any point S in the first focal plane (Fig. 48), and through it draw SJ_1J_2 parallel to the axis; join J_2F''. All light proceeding from S will emerge from the system in a direction parallel to J_2F'', *e.g.* SI_1I_2R. From S draw the ray SD_1K' parallel to J_2F'', representing such a ray, and from D_2 the point on the second principal plane corresponding to D_1, draw $D_2K''E$ parallel to J_2F'', cutting the axis in K″. Then K′ and K″ are the first and second nodal points of the system. We shall see the use of them in the next section. Meanwhile we will devote a little attention to the diagram. We notice first that every ray incident on the first principal plane travels parallel to the axis until it meets the second principal plane; this is a characteristic property of these planes, and as any object in one plane, when viewed from the other side, will be seen without any alteration in size, they are often called planes of unit magnification, or the Unit Planes.

As the sides of the $\triangle F'SK'$ are parallel to the sides of the $\triangle H''J_2F''$, and as $F'S = H''J_2$,

$$K'F' = F''H'' \text{ or } F''.$$

Also the $\triangle EF''K'' = \triangle D_1J_1S$, as the sides are parallel, and as $EF'' = D_2J_2 = D_1J_1$,

$$\therefore K''F'' = SJ_1 = F'H' \text{ or } F'$$

It is also clear that $K'K'' = D_1D_2 = H'H''$,
and that $K''H'' = K'H' = F''H'' + F'H' = F'' + F'$.

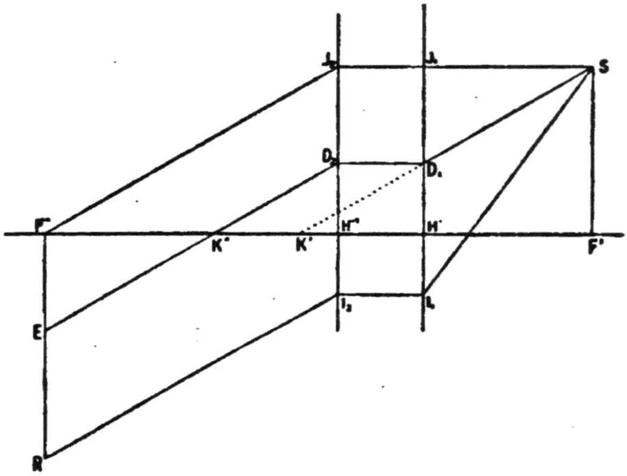

Fig. 48.

We have found then all the cardinal points of the standard emmetropic eye which are given in the subjoined table.

Cardinal Points of the Emmetropic Eye.

$H'A_0$	− 1·789 mm	$H'H''$, $K'K''$	0·353 mm.
$H''A_0$	− 2·142 mm.		
$K'A_0$	− 7·04 mm.	$H'K'$, $H''K''$	5·25 mm.
$K''A_0$	− 7·39 mm.		
$F'A_0$	13·75 mm.	$F'H''$, $K''F''$	15·54 mm.
$F''A_0$	− 22·93 mm.		
		$F''H''$, $K'F'$	−20·79 mm.

In every system in which the initial and the final media have the same refractive index (*i.e.* when $F''H'' = H'F'$) the nodal points K', K'' coincide with the principal points H', H''. We see also that if we regard the principal points as coincident in the human eye in H, the nodal points coincide too in one point K, and the distance of this nodal point to the principal point, or KH, $= F''H'' + F'H' = -20{\cdot}79 + 15{\cdot}54 = -5{\cdot}25$ mm.

Geometrical Construction of the Image.—To find the image of AB (Fig. 49), we first mark the position of the cardinal points on the axis. We then join AK′, and draw K″*a* parallel to it. Then we can either draw AJ″ parallel to the axis and J″F″*a* through F″, or we can draw AF′I′ through F′, and I′I″*a* parallel to the axis. In either case, *a* is the point of intersection of the line with that drawn through the nodal point.

Note that $\frac{i}{o} = \frac{H'I'}{BA} = \frac{F'H'}{F'B} = \frac{F'H'}{F'H' - BH'} = \frac{F'}{F' - p}$

and that $\frac{i}{o} = \frac{ba}{H''J''} = \frac{F''b}{F''H''} = \frac{F''H'' - bH''}{F''H''} = \frac{F'' - q}{F''}$

On p. 75 it was shown that when thin lenses are considered $\frac{i}{o} = \frac{q}{p}$, and we see from Fig. 49 that $\frac{i}{o} = \frac{bK''}{BK'}$ so that the nodal points of a complex system play the part of the optical centre (O) in a thin lens. The distances *b*K″ and BK′ are usually denoted by g'' and g' in the books.

In the simplified schematic eye described above, the nodal point K acts as if it were the centre of a convex spherical surface of radius $-5{\cdot}25$ mm. that forms the boundary of a medium of refractive index—

$$-\frac{F''}{F'} \quad \text{or} \quad \frac{20{\cdot}79}{15{\cdot}54} \approx 1{\cdot}338$$

The tangent of the visual angle subtended at the nodal point of the eye (p. 77) by an object is the same as the tangent of the angle subtended by the retinal image at K. If this retinal image be *i* mm. in height,

$$\tan\theta = \frac{i}{KF''} = \frac{i}{F'} = \frac{i}{15{\cdot}54}$$

It is often simpler, when the object is very distant, to consider an optical instrument and the eye as forming one complex system, as in that case $\tan\theta$ is practically constant, and if K″ denote the second Nodal Point of the system and K that of the unaided eye, it is clear that the magnification of the system is as *b*K″ : *b*K.

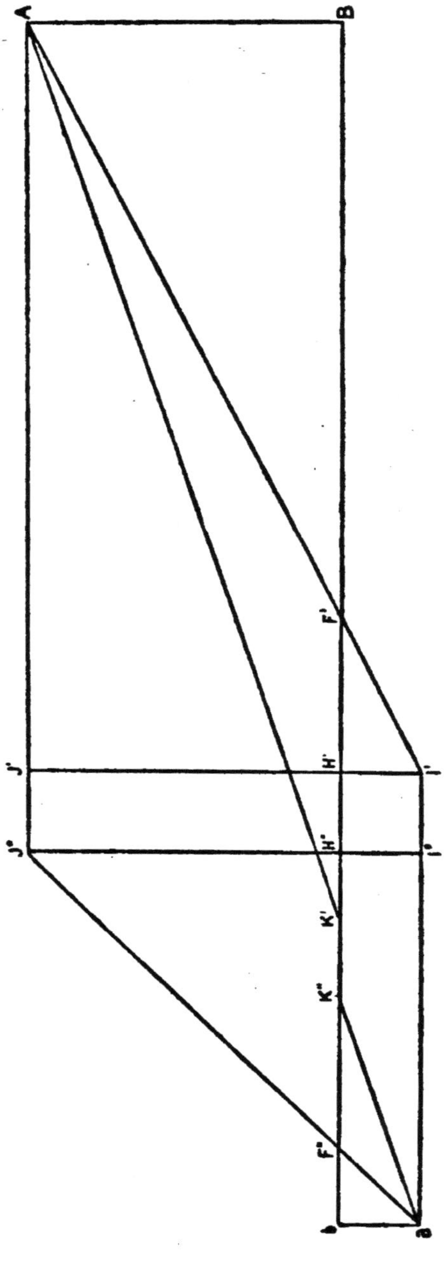

FIG. 49.

As an example, we may explain the action of a Baden Powell lens, which is used as a handy pocket opera glass. It is simply a convex lens of weak power which is held at arm's length, and distant objects are viewed through it; these then appear larger to the eye, or, when less correctly expressed, they are said to seem nearer.

We will suppose that the glass is $+0{\cdot}5$D where $f' = 2000$ mm., and that it is held at a distance of 500 mm., or about 20 inches, from the eye.

On considering the lens and the eye as one complex system, and finding the cardinal points, we get the following values:—

$$K = t + f'' - F' = 500 - 2000 - 15{\cdot}54 = -1515{\cdot}54$$

$$\therefore h' \text{ or } \frac{f't}{K} = \frac{2000 \times 500}{-1515{\cdot}54} = -659{\cdot}83 \text{ mm.}$$

$$h'' \text{ or } \frac{F''t}{K} = \frac{-20{\cdot}79 \times 500}{-1515{\cdot}54} = 6{\cdot}859 \text{ mm.}$$

$$F' = \frac{-f'F'}{K} = \frac{-2000 \times 15{\cdot}54}{-1515{\cdot}54} \approx 20{\cdot}51$$

$$F'' = \frac{f''F''}{K} = \frac{-2000(-20{\cdot}79)}{-1515{\cdot}54} \approx -27{\cdot}436$$

It is clear, then, that the first principal point and the first focus are both more than 100 mm. behind the eye. This is of no importance in the problem now before us; but the second principal focus must be situated on the retina if the object is to be seen distinctly, so we must find whether this is so.

F$''$ is situated 27·436 mm. behind H$''$, which is 6·859 mm. in front of the eye, consequently **F**$''$ is 20·577 mm. behind the principal plane of the eye. As in the standard eye F$''$ is 20·79 mm. behind its principal plane, we see that, unless the eye be 0·213 mm. too short, a sharp image will not be formed on the retina.

Now, few eyes are of exactly the standard length, and we will suppose that the eye considered is too short by at least this small amount (less than 0·75D of hypermetropia), so that b**K**$'' =$ **F**$''$**K**$'' = -F'$.

The magnification (M) will be as F″K″ : F″K

or
$$M = \frac{F''K''}{F''K} = \frac{-F'}{-F''} = \frac{20{\cdot}51}{15{\cdot}54} \approx 1{\cdot}32$$

Consequently, the distant object will appear larger by nearly a third. If the eye be myopic or too long, the Baden Powell lens is useless without some contrivance to make the image definite. The simplest effective contrivance is a card with a pinhole in it held close to the eye. In this way the circles of confusion on the retina are made much smaller, so that the image may be regarded as sharply defined.

Graphic Method for Cardinal Points.—Professor Sampson has devised a most ingenious method of finding the cardinal points of a thick lens by a graphic method. It is an extension of the graphic methods which we have frequently illustrated in the preceding sections.

The diagram (Fig. 50) shows this method applied to the case of the meniscus we discussed on p. 95. The thickness of the lens AB is 3, so we measure AB in the positive direction (upwards) 3 units. As A indicates the first surface, we mark off $f_1'A$ in the negative direction to represent -8 units, and $f_1''A$ in the positive direction to represent 12 units. Similarly, dealing with the second surface B, we make $f_2'B$ equal to $+6$ units, and $f_2''B$ equal to -4 units. The parallelogram referring to the first surface (A) is completed at K_1, and that referring to B is completed at K_2.

Join K_1K_2, and produce to meet the horizontal line through B at H″, cutting the horizontal line through A at H′.

Then H′A is h', the distance of the first surface from H′, and H″B is h'', the distance of the second surface from H″. In this case they are both measured from right to left, so they are both negative: $H'A = -2\frac{2}{3}$, and $H''B = -1\frac{1}{3}$.

Join K_1 and f_2', and produce to meet the horizontal line through A in F′; and join K_2f_1'', and produce to meet the horizontal line through B in F″.

Then F′H′ is positive and is found to measure $5\frac{1}{3}$ units, while F″H″ is found to measure $-5\frac{1}{3}$ units. It will be found that F′H′ and F″H″ correctly indicate the two principal focal distances of the thick lens illustrated in Fig. 44.

The method is of delightful simplicity and presents no difficulty whatever, if due attention be paid to the directions of the positive and negative measurements, and if it be noted that the points f_1'' and f_2' must be in the same straight line

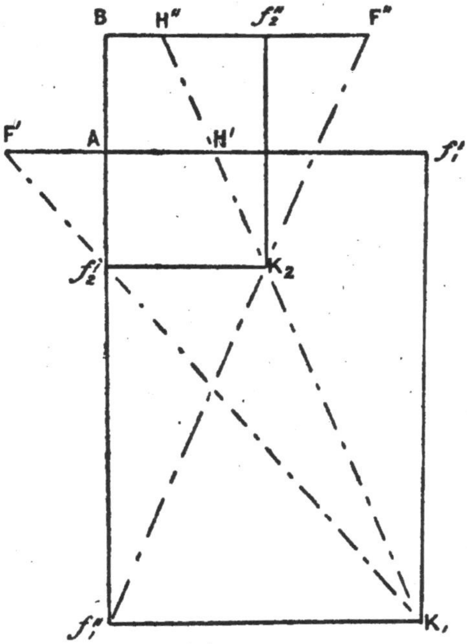

FIG. 50.

as AB, and that K_1 is joined to f_2' (subscript 2), while K_2 is joined to f_1'' (subscript 1).

To show the generality of the construction, it will be sufficient here to notice that

$$f_1''f_2' = f_1''A + AB - f_2'B = t + f_1'' - f_2' = K$$

and that from the similar triangles in the figure

$$\frac{F'A}{K_1f_1''} = \frac{f_2'A}{f_2'f_1''} \quad \text{or} \quad \frac{F'H' + H'A}{f_1'A} = \frac{f_2'B - AB}{-f_1''f_2'}$$

or
$$\frac{F' + h'}{f_1'} = \frac{t - f_2'}{K} \quad \text{(Cf. p. 94)}$$

Similarly, $\frac{F''B}{K_2 f_2'} = \frac{f_1''B}{f_1''f_2'}$ or $\frac{F''H'' + H''B}{f_2''B} = \frac{f_1''A + AB}{f_1''f_2'}$

or $$\frac{F'' + h''}{f_2''} = \frac{f_1'' + t}{K} \quad \text{(Cf. p. 94)}$$

(The above is not given as a proof of the construction, for clearly F′A = F′K′ + K′A, and F″B = F″K″ + K″B, or indeed the sum of any other lines with the same end points. It will be found that if the final medium has a different refractive index from that of the initial medium, the points marked H′ and H″ really denote the two nodal points K′ and K″ of the system. As from p. 101 we know that when $F'' = -F'$, the nodal points coincide with the principal points, we may regard the points H′ and H″ determined by this construction as always denoting the nodal points.)

QUESTIONS.

(1) The focal length of a convex lens is 6 ins.; an object is placed 36 ins. from it. What is the relative size of the image, and where is it formed?

(2) Show that $f'f'' = (f' - p)(f'' - q)$ in all cases.

(3) The radius of curvature of the first surface of a thin converging lens is −6 ins. If its focal length be 10 ins., and if the index of refraction be 1·52, what is the radius of curvature of the other surface? What would its focal length be when placed in a tank of water ($\mu = \frac{4}{3}$)?

(4) A convex lens of focal length $\frac{1}{5}$ in. is used as a magnifying glass. The nearest point of distinct vision is 10 ins. from the nodal point of the eye. Find the magnifying power (i.) when the lens is $\frac{1}{2}$ in. from the nodal point, (ii.) when it is $1\frac{1}{4}$ in. from the nodal point; and (iii.) when the object is $\frac{1}{5}$ in. from the lens, and the eye is adapted for distance.

(5) The cardinal points of the following thick biconvex lens are required where $r_1 = -4$, $r_2 = 2$, $t = 3$, and $\mu = 1·5$. Use the graphic method, and test your results by the algebraic formulæ for the position of the cardinal points.

APPENDIX

The following notes are intended for those who require some further knowledge of the subject which the previous elementary treatment rendered impossible. At the same time care has been taken to include nothing which has no practical application; academic points of merely mathematical interest have been rigidly excluded.

Oblique Reflection. Distances of the Focal Lines.—We will now give the proof of the formula quoted on p. 33, denoting

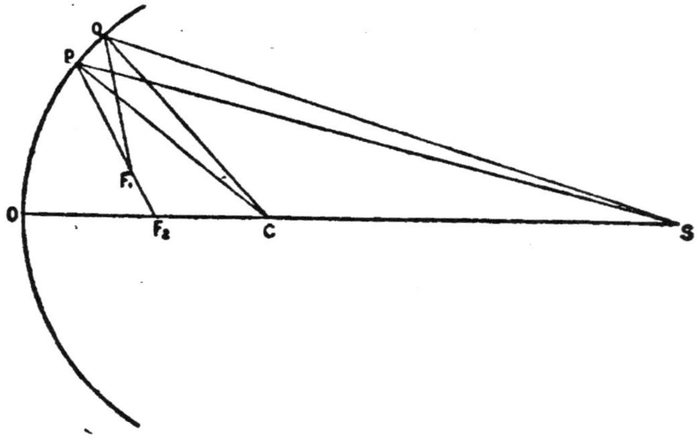

Fig. 51.

SP by u and F_1P and F_2P by v_1 and v_2, the distances of the focal lines due to the thin incident pencil PSQ (Fig. 51).

$$\begin{aligned} PCQ &= OCQ - OCP = CSQ + SQC - (CSP + SPC) \\ &= CSQ - CSP + \tfrac{1}{2}SQF_1 - \tfrac{1}{2}SPF_1 \\ 2PCQ &= 2PSQ + SQF_1 - SPF_1 \quad . \quad . \quad . \quad . \quad . \quad . \quad . \quad . \quad . \quad (a) \end{aligned}$$

But owing to the equality of the angles formed by the intersecting lines SP and F_1Q, $PSQ + SQF_1 = PF_1Q + SPF_1$, and on substituting this expression in (*a*) we get

$$2PCQ = PSQ + PF_1Q \quad . \quad . \quad . \quad . \quad . \quad (b)$$

Now, as the formula that we wish to find is only true for thin pencils PSQ and PF_1Q, we may substitute the chord PQ for the arc PQ, and regard PSQ and PF_1Q as triangles; moreover, CQS or $F_1QC = \phi$.

Then in $\triangle$ PSQ, $\dfrac{PQ}{SP} = \dfrac{\sin PSQ}{\sin PQS} = \dfrac{\sin PSQ}{\sin(90° + CQS)}$

$$\therefore \frac{PQ}{SP} = \frac{\sin PSQ}{\cos\phi}$$

and in $\triangle PF_1Q$, $\dfrac{PQ}{F_1P} = \dfrac{\sin PF_1Q}{\sin PQF_1} = \dfrac{\sin PF_1Q}{\sin(90° - F_1QC)}$

$$\therefore \frac{PQ}{F_1P} = \frac{\sin PF_1Q}{\cos\phi}$$

therefore in the limit when PSQ and PF_1Q are very small

$$PSQ = PQ\frac{\cos\phi}{u},\ PF_1Q = PQ\frac{\cos\phi}{v_1} \text{ and } PCQ = \frac{PQ}{r} \quad . \quad (c)$$

Note that as PQ is measured from below upwards, but SP from right to left, the angles PSQ and PQS are measured in reverse directions; a similar precaution must be exercised in dealing with the angles PF_1Q and PQF_1.

On substituting the expressions in (*c*) for those in (*b*), we can write

$$\frac{2PQ}{r} = PQ\frac{\cos\phi}{u} + PQ\frac{\cos\phi}{v_1}$$

or

$$\frac{1}{u} + \frac{1}{v_1} = \frac{2}{r\cos\phi} \quad . \quad . \quad . \quad . \quad . \quad . \quad . \quad . \quad . \quad (A)$$

Again, since $\triangle F_2PC + \triangle CPS = \triangle F_2PS$

$$\tfrac{1}{2}v_2r\sin\phi + \tfrac{1}{2}ru\sin\phi = \tfrac{1}{2}v_2u\sin 2\phi$$

or, on dividing throughout by $\frac{1}{2}v_2ru\sin\phi$,

$$\frac{1}{u} + \frac{1}{v_2} = \frac{\sin 2\phi}{r\sin\phi} = \frac{2\sin\phi\cos\phi}{r\sin\phi} = \frac{2\cos\phi}{r} \quad . \quad . \quad (B)$$

The reader may be inclined to think that the consideration of such very thin pencils is of little practical use, but it must be remembered that only very thin pencils can enter the pupil of the eye, so that only thin pencils need be considered, if the image formed by the mirror is viewed directly by the eye.

Camera Lucida.—The Wollaston prism shown in Fig. 52 is the usual form of camera lucida that is used in sketching objects from nature. It is a glass prism that presents four angles, one of which is 90°, the opposite angle 135°, and the remaining two angles are each 67½°. Light from the object enters one of the faces normally, and is transmitted without deviation to the second face as SI; at I, however, its angle of incidence is 67½°, much greater than the critical angle for glass, so that it is totally reflected at I as IR; again at R it undergoes total reflection; and finally it emerges normally to the upper surface towards the eye at E.

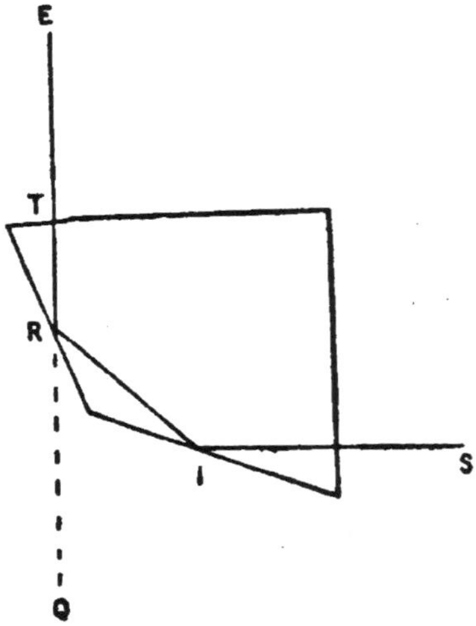

FIG. 52.

The eye will therefore see a virtual image of the object projected downwards in the direction EQ. In practice the eye is placed over the edge near T, so that one half of the pupil receives the light from the prism, while the other half is viewing the sketching-block and pencil placed below in the neighbourhood of Q. It is most important that the image of the object should be accurately projected on to the plane of the sketching-block, so the upper border at T is ground concave so that it will have the effect of a −4D lens. In this way the image is projected about 10 ins. from the prism, which will be a convenient distance for the sketch to be made. The image seen is erect, as there are two reflections; had there been only one reflection the image would have been upside down.

Oblique Refraction at Plane Surface. Focal Lines.—We will now consider the refraction of a thin oblique pencil at a plane surface. If POQ (Fig. 53) be an incident oblique pencil originating from the source O in a rare medium and refracted at PQ in the direction RR′, the primary and secondary (virtual) focal lines will be formed at F_1 and F_2. As the pencil considered is very thin, we may substitute the angles POQ and PF_1Q for sin POQ and sin PF_1Q, and we may disregard the difference between the angles AOP and AOQ and consider that they are

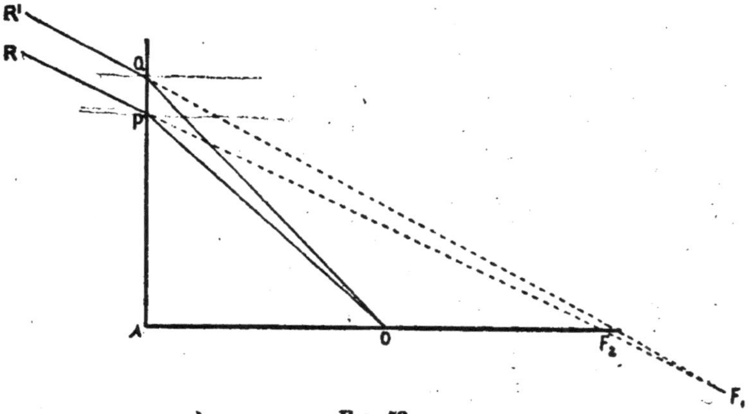

FIG. 53.

each of them equal to ϕ, and similarly we may consider F_2 as a point, and consider that $AF_2P = AF_2Q = \phi'$.

If we denote OP by u, F_1P by v_1, and F_2P by v_2, we have

in $\triangle$ POQ, $$\frac{PQ}{OP} = \frac{\sin POQ}{\sin PQO} = \frac{\sin POQ}{\cos AOQ}$$

and in $\triangle PF_1Q$, $$\frac{PQ}{F_1P} = \frac{\sin PF_1Q}{\sin PQF_1} = \frac{\sin PF_1Q}{\cos AF_2Q}$$

so $$POQ = \frac{PQ}{u}\cos\phi \quad \text{and } PF_1Q = \frac{PQ}{v_1}\cos\phi'.$$

Now, $\mu_0 \sin\phi = \mu' \sin\phi'$,

$$\therefore \quad \frac{d\phi}{d\phi'} = \frac{\mu' \cos\phi'}{\mu_0 \cos\phi} \quad . \quad . \quad . \quad . \quad . \quad . \quad . \quad (a)$$

and we wish to find from Fig. 53 an expression for $\frac{d\phi}{d\phi'}$, or the limiting value of $\frac{\Delta\phi}{\Delta\phi'}$, *i.e.* $\frac{POQ}{PF_1Q}$.

When, therefore, POQ and PF_1Q become infinitesimal,

$$\frac{d\phi}{d\phi'} = \frac{\frac{PQ}{u}\cos\phi}{\frac{PQ}{v_1}\cos\phi'} \text{ or } \frac{v_1\cos\phi}{u\cos\phi'} = \frac{\mu'}{\mu_0}\cdot\frac{\cos\phi'}{\cos\phi} \text{ from } (a)$$

$$\therefore\ v_1 = \frac{\mu'}{\mu_0}u\cdot\frac{\cos^2\phi'}{\cos^2\phi} \quad . \quad . \quad . \quad . \quad . \quad . \quad . \quad . \quad . \quad . \quad . \quad (A)$$

And since $\sin\phi = \frac{AP}{OP}$ and $\sin\phi' = \frac{AP}{F_2P}$,

$$\frac{\mu'}{\mu_0} \text{ or } \frac{\sin\phi}{\sin\phi'} = \frac{F_2P}{OP} = \frac{v_2}{u}$$

$$\therefore\ v_2 = \frac{\mu'}{\mu_0}u \quad . \quad . \quad . \quad . \quad . \quad . \quad (B)$$

The refracted pencil will be astigmatic, and a sphenoid will be formed between the secondary and primary focal lines (F_2 and F_1) exactly like that described on p. 65, when oblique refraction at a concave surface was considered from a source of light whose distance u was less than μ (F_1P). A blurred image of the point O will be formed at the position of the "circle of least confusion" between F_2 and F_1 (as represented by D in Fig. 56, where, however, F_2P is greater than F_1P).

Oblique Refraction through a Plate.—Let t denote the thickness of the plate, and let l denote the length of the path NM of the thin pencil through the plate in Fig. 19. For the first refraction we have $v_1 \frac{\mu'}{\mu_0}u\cdot\frac{\cos^2\phi'}{\cos^2\phi}$ and $v_2 = \frac{\mu'}{\mu_0}u$ by (A) and (B). For the second refraction at M the angle of incidence is ϕ', and that of emergence is ϕ, while the relative index is $\frac{\mu_0}{\mu'}$, so that on replacing u by $v_1 + l$ or $v_2 + l$ in the respective equations, and making the other appropriate substitutions, we get

$$V_1 = \frac{\mu_0}{\mu'}(v_1 + l)\frac{\cos^2\phi}{\cos^2\phi'} \quad \text{or} \quad u + \frac{\mu_0}{\mu'}l\frac{\cos^2\phi}{\cos^2\phi'} \quad . \quad (A')$$

$$V_2 = \frac{\mu_0}{\mu'}(v_2 + l) \quad \text{or} \quad u + \frac{\mu_0}{\mu'}l \quad . \quad . \quad . \quad . \quad . \quad (B')$$

The distance between the two foci, $V_2 - V_1$ or $\frac{\mu_0}{\mu'}l\left(1 - \frac{\cos^2\phi}{\cos^2\phi'}\right)$

may be taken as a measure of the indistinctness of the image. However oblique the pencil the distance separating the two foci cannot exceed $\frac{\mu_0}{\mu'}l$; for direct pencils of course $\phi = 0$, and the two foci are coincident at one and the same point distant $u + \frac{\mu_0}{\mu'}l$ from the distal surface of the plate. If preferred we may substitute $t \sec \phi'$ for l, but the form given above is easier to remember.

When a small object S is seen obliquely through a glass plate (Fig. 54) the following points may be observed:—

FIG. 54.

(1) The image S′ is blurred, as it is represented by the "circle of least confusion" between the two focal lines.

(2) The upward displacement is greater than when the object is viewed normally, for in that case, as we found on p. 40, the upward displacement is $\frac{\mu - 1}{\mu}t$, or one-third the thickness of the glass plate.

(3) There is in addition a lateral displacement.

Refraction at Spherical Surface. Focal Lines.—The length of v_1 is determined in this way (Fig. 55). When the pencil is thin, CPO or CQO may be represented by ϕ, and CPF_1 or CQF by ϕ'.

In $\triangle$ OPQ, $$\frac{PQ}{OP} = \frac{\sin POQ}{\sin PQO} = \frac{\sin POQ}{\sin (90 - OQC)} = \frac{\sin POQ}{\cos CQO}$$

Similarly, in $\triangle$ F_1PQ,

$$\frac{PQ}{F_1P} = \frac{\sin PF_1Q}{\sin PQF_1} = \frac{\sin PF_1Q}{\sin (90 - F_1QC)} = \frac{\sin PF_1Q}{\cos CQF_1}$$

Therefore in the limit when POQ and PF_1Q are very small,

$$POQ = PQ\frac{\cos \phi}{u}, \quad PF_1Q = PQ\frac{\cos \phi'}{v_1} \quad \text{and} \quad PCQ = \frac{PQ}{r}$$

Now, $\mu_0 \sin \phi = \mu' \sin \phi'$,

$$\therefore \frac{d\phi}{d\phi'} = \frac{\mu'}{\mu_0} \cdot \frac{\cos \phi'}{\cos \phi}$$

As we wish to find from the figure an expression for $\frac{d\phi}{d\phi'}$, or the limiting value of $\frac{\Delta\phi}{\Delta\phi'}$, we must no longer disregard the difference between CQO and CPO and that between CQF_1 and CPF_1. Since the angles at the intersection L are equal,

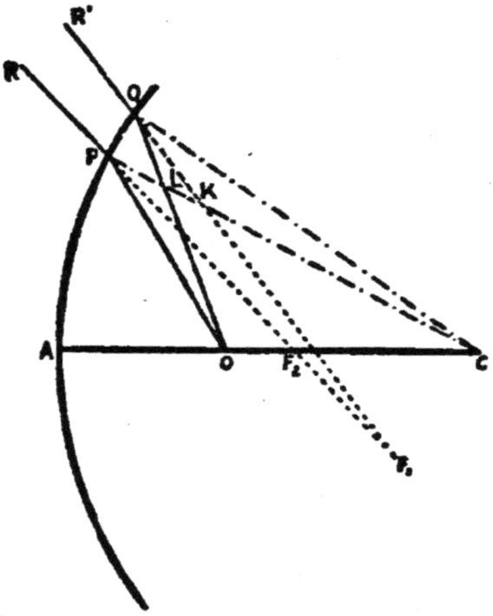

FIG. 55.

$$POL + LPO = LCQ + CQL$$

or
$$POQ + \phi = PCQ + \phi + \Delta\phi$$

And similarly, since the angles at K are equal,

$$PF_1K + KPF_1 = KCQ + CQK$$

or
$$PF_1K + \phi' = PCQ + \phi' + \Delta\phi'$$

$$\therefore \frac{\Delta\phi}{\Delta\phi'} = \frac{POQ - PCQ}{PF_1K - PCQ}$$

or in the limit
$$\frac{d\phi}{d\phi'} = \frac{PQ\left(\frac{\cos\phi}{u} - \frac{1}{r}\right)}{PQ\left(\frac{\cos\phi'}{v_1} - \frac{1}{r}\right)}$$

But
$$\frac{d\phi}{d\phi'} = \frac{\mu' \cos\phi'}{u_0 \cos\phi}$$

$$\therefore \frac{\mu' \cos^2 \phi'}{\mu_0 v_1} - \frac{\mu' \cos \phi'}{\mu_0 r} = \frac{\cos^2 \phi}{u} - \frac{\cos \phi}{r}$$

$$\therefore \frac{\mu' \cos^2 \phi'}{\mu_0 v_1} - \frac{\cos^2 \phi}{u} = \frac{1}{r}\left(\frac{\mu'}{\mu_0} \cos \phi' - \cos \phi\right)$$

$$= \frac{\sin \phi \cos \phi' - \cos \phi \sin \phi'}{r \sin \phi'}$$

or

$$\frac{\mu' \cos^2 \phi'}{\mu_0 v_1} - \frac{\cos^2 \phi}{u} = \frac{\sin (\phi - \phi')}{r \sin \phi'} \quad . \quad . \quad . \quad . \quad \text{(A)}$$

Again, since

$$\triangle \text{CPO} = \triangle \text{CPF}_2 + \triangle \text{F}_2\text{PO}$$

$$\tfrac{1}{2} ru \sin \phi = \tfrac{1}{2} rv_2 \sin \phi' + \tfrac{1}{2} v_2 u \sin (\phi - \phi')$$

or on dividing throughout by $\frac{1}{2} ruv_2 \sin \phi'$

$$\frac{\sin \phi}{\sin \phi'} \cdot \frac{1}{v_2} = \frac{1}{u} + \frac{\sin (\phi - \phi')}{r \sin \phi'}$$

$$\text{or} \quad \frac{\mu'}{\mu_0 v_2} - \frac{1}{u} = \frac{\sin (\phi - \phi')}{r \sin \phi'} \quad . \quad . \quad . \quad . \quad . \quad \text{(B)}$$

These formulæ (A) and (B) are only true for extremely thin pencils, but it is only such that can enter the pupil of the eye, so that they are applicable in all the optical questions relating to oblique vision in the human eye. They will also be used in the next section when we deal with thin oblique centric pencils traversing a lens.

It is evident from formulæ (A) and (B) that

$$\frac{\mu_1}{\mu_0 v_1}(1 - \sin^2 \phi') - \frac{1 - \sin^2 \phi}{u} = \frac{\mu'}{\mu_0 v_2} - \frac{1}{u}$$

or

$$\frac{\mu'}{\mu_0}\left(\frac{1}{v_1} - \frac{1}{v_2}\right) = \frac{\mu'}{\mu_0} \cdot \frac{\sin^2 \phi'}{v_1} - \frac{\sin^2 \phi}{u} = \sin^2 \phi\left(\frac{\mu_0}{\mu_1 v_1} - \frac{1}{u}\right)$$

$$\therefore \frac{\mu'}{\mu_0}(v_2 - v_1) = \frac{v_2 \sin^2 \phi}{\mu' u}(\mu_0 u - \mu' v_1).$$

When μ_0 has the value 1, this expression takes the form

$$v_2 - v_1 = \frac{v_2 \sin^2 \phi}{\mu^2 u}(u - \mu v_1),$$

and we see that when v_2 and v_1 carry the same sign, *i.e.* when the two focal lines are on the same side of the refracting surface,

$$v_2 \gtreqless v_1 \quad \text{as} \quad u \gtreqless \mu v_1$$

On considering the caustic curve formed by a wide divergent pencil issuing from a point source, it is clear that the cusp of the caustic points away from the spherical surface when $v_2 > v_1$, but that it points towards the surface when $v_2 < v_1$. No caustic is formed when $v_2 = v_1$; this can only occur when $\phi = 0$, or when $u = \mu v_1$; the refracting surface must then be concave, for if $\phi = 0$, $u = r$, and if $u = \mu v_1$, as v_1 must be positive, it must be greater or less than u, according as μ is less or greater than 1.

Circle of Least Confusion.—It has been shown that if the surface be not aplanatic for the source, whenever the incident pencil is oblique the refracted or reflected pencil is astigmatic, crossing itself in two focal lines and tracing out a sphenoid surface between them. It is required to find the position and size of the smallest cross-section of this sphenoid surface represented by DK in Fig. 56, which is practically a repetition of Fig. 15. The surface (ab) at P is that part of the mirror or lens which gives rise to the astigmatic pencil that has its focal lines at F_1 and F_2. For geometrical convenience squares have been described on the radius (k) of the circle of least confusion, and on the radius R of the effective aperture of the receiving instrument. Now, DP (or x) and k, *i.e.* the distance and side of the small square at D, can be easily expressed in terms of a and b. Note that when comparing the vertical sides of the rectangles at P and D, while a is measured upwards k is measured downwards, so that they carry opposite signs.

$$\frac{F_1P}{a} = \frac{F_1D}{-k} \quad \text{or} \quad \frac{DF_1}{k} \qquad \therefore \frac{v_1}{a} = \frac{x - v_1}{k} \quad . \quad . \quad (1')$$

$$\frac{F_2P}{b} = \frac{F_2D}{k} \qquad \therefore \frac{v_2}{b} = \frac{v_2 - x}{k} \quad . \quad . \quad . \quad . \quad . \quad (2')$$

$$\therefore \frac{v_1}{a} + \frac{v_2}{b} = \frac{v_2 - v_1}{k} \quad . \quad . \quad . \quad . \quad . \quad . \quad (a')$$

On eliminating k from (1′) and (2′) we obtain

$$a\frac{x - v_1}{v_1} = b\frac{v_2 - x}{v_2}$$

which, on dividing by x, reduces to

$$\frac{a}{v_1} + \frac{b}{v_2} = \frac{a + b}{x} \quad . \quad . \quad . \quad . \quad . \quad . \quad . \quad (b')$$

In practice, however, we have more often to consider the cases in which a real image is formed at the back of an eye or at the back of a photographic camera. It is, therefore, much more convenient to express the distance (x) and the radius (k) of the circle of least confusion in terms of the distance (s) of the receiving instrument and the radius (R) of the stop used with it.

Let H′ represent the first principal plane of the receiving instrument (distant H′P or s), which only allows a pencil of radius R to be transmitted through it; if the receiving instrument be an eye, R represents the radius of an equivalent pupil placed in the first principal plane of the eye.

By similar triangles we have

$$\frac{F_1H'}{R} = \frac{F_1D}{-k} \quad \text{or} \quad \frac{DF_1}{k} \qquad \therefore \frac{v_1 - s}{R} = \frac{x - v_1}{k}. \quad . \quad (1)$$

and
$$\frac{F_2H}{R} = \frac{F_2D}{k} \qquad \therefore \frac{v_2 - s}{R} = \frac{v_2 - x}{k} \quad . \quad . \quad . \quad . \quad . \quad . \quad (2)$$

$$\therefore \frac{v_1 + v_2 - 2s}{R} = \frac{v_2 - v_1}{k}$$

or
$$k = \frac{R(v_2 - v_1)}{v_1 + v_2 - 2s} \quad . \quad . \quad . \quad (a)$$

Again, from (1) and (2) we also see that

$$\frac{k}{R} = \frac{x - v_1}{v_1 - s} = \frac{v_2 - x}{v_2 - s}$$

or
$$x(v_2 - s) - v_1v_2 + v_1s = v_1v_2 - v_2s - x(v_1 - s)$$

$$\therefore x(v_1 + v_2 - 2s) = 2v_1v_2 - s(v_1 + v_2)$$

or
$$x = \frac{2v_1v_2 - s(v_1 + v_2)}{v_1 + v_2 - 2s}. \quad . \quad . \quad (b)$$

These two equations (a) and (b), determining the radius and the distance of the circle of least confusion of an oblique pencil, are always true whether reflection or refraction are considered.

If the receiving instrument be an eye, and the lens at P be a spectacle glass worn in its appropriate place (the first focal plane of the eye), a simple expression can be obtained for the radius r of the retinal confusion circle formed by the retinal image of this confusion circle at D.

For
$$\frac{r}{k} = \frac{F'}{F' - p}$$

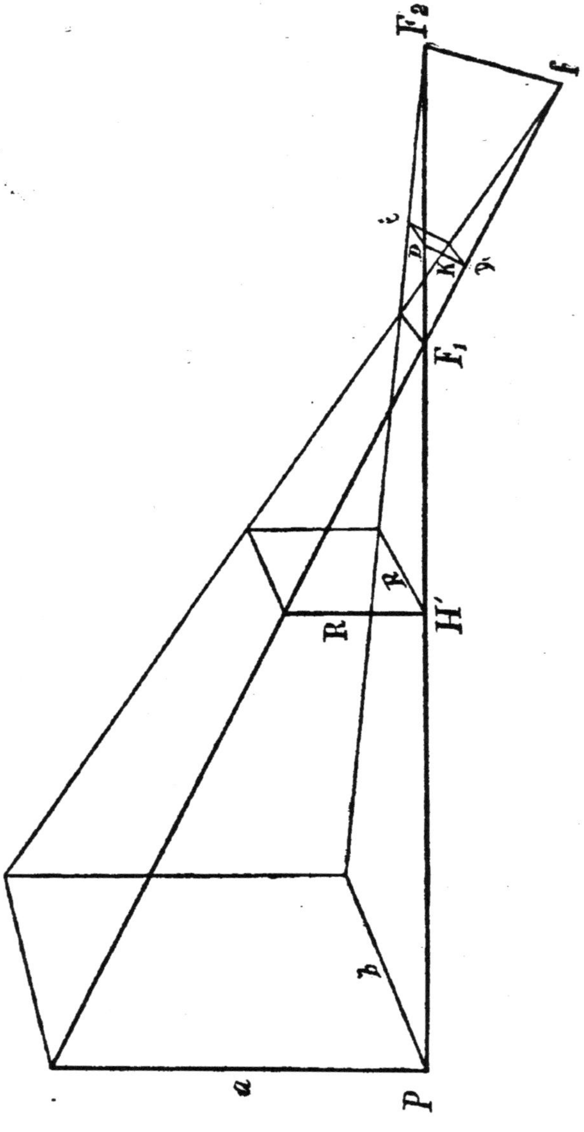

Fig. 56.

But as the lens is placed in the first focal plane, PH′ = F', so $s = -F'$, and $F' - p$ = PH′ − DH′ = PD = $-x$;

$$\therefore\ r = \frac{sk}{x} = \frac{sR(v_2 - v_1)}{2v_1v_2 - s(v_1 + v_2)} \quad \text{or} \quad \frac{F'R(v_1 - v_2)}{2v_1v_2 + F'(v_1 + v_2)}$$

Least Circle of Aberration.—These formulæ are of no use when we are considering the spherical aberration of a lens used with full aperture, as in Fig. 42. Let y (Fig. 57) denote

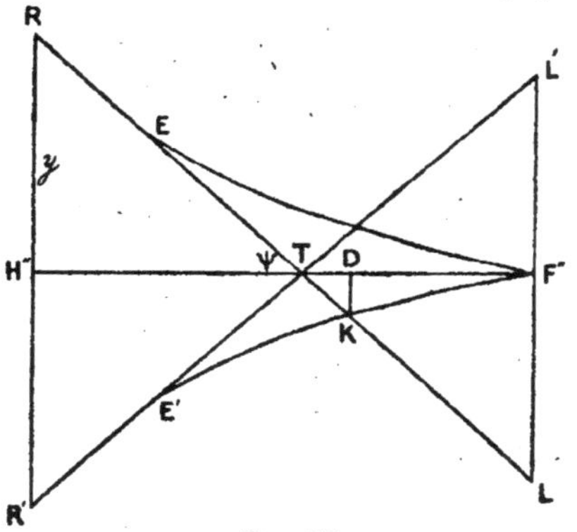

Fig. 57.

the semi-aperture of a convex lens the focus of which is at F″, and let RTL denote an extreme ray; it is clear that if a screen were placed at F″, there would be a bright point at F″ which would be surrounded by a halo of light extending to L and L′. The line F″L is called the Lateral Aberration (l) of the extreme ray RTL, while F″T or a is called the Longitudinal Aberration of RTL which cuts the axis at an angle H″TR or ψ.

Clearly $\quad a$ = F″H″ − TH″ = $f'' - y \cot \psi$,

and $\quad l = -a \tan \psi$.

The point E represents the position of the first focal line (f_1), and the point T that of the second focal line (f_2) of the extreme ray RTL, and the curve EF″E′ represents the Caustic Curve formed by the lens when its effective aperture is R′R. It is obvious that the light is concentrated over the smallest area at D, which marks the site of the Least Circle of Aberration, its radius DK or k is determined by the point where the extreme ray RTL cuts the caustic E′F″. It is usually stated that

F″D = $\frac{3}{4}a$, and that $k = \frac{1}{4}l$.

This is only a very rough approximation, as the position of D can only be determined by tracing the caustic,[1] which may be of almost any shape, some being long and narrow, while others are short and stumpy, so that no general expression can be given for the position of D. If in any case the position of D is given, of course $k = \text{TD} \tan \psi$.

Thin Lens. Thin Oblique Centric Pencils.—It has been already demonstrated that all oblique pencils which traverse the optical centre of a lens emerge at an angle equal to that of incidence (ϕ) at the first surface, and that the angle of incidence at the second surface is equal to the angle of refraction (ϕ') at the first surface. This fact is universally true, however different the radii of curvatures of the respective surface may be.

Fig. 58 gives in a simple diagrammatic way the results of the refraction of a thin oblique pencil PO that traverses the centre O of a thin concave lens represented by the plane HO.

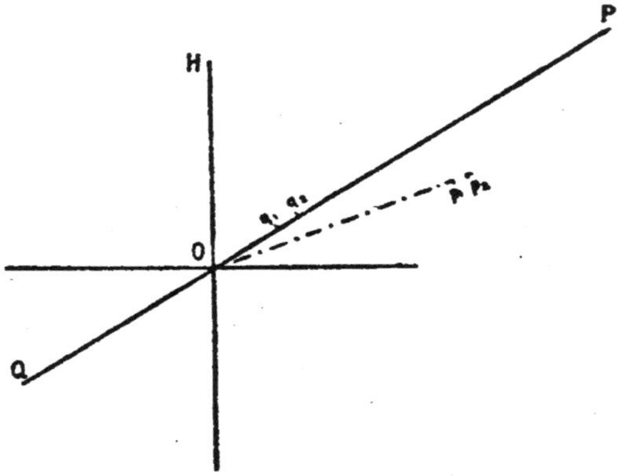

FIG. 58.

By the method of p. 116 we find the positions of p_1 and p_2, the primary and secondary focal lines due to refraction at the first surface. Then, regarding p_1 as the virtual object for the second surface, we find q_1 as the final primary focal line due to this oblique centric refraction through the lens, and similarly q_2 represents the final secondary focal line.

[1] A paper of mine in the *Proceedings* of the Optical Convention, 1912, describes a method of tracing Caustic Curves.

Hence, if r_1 and r_2 represent the radii of the first and second surfaces respectively, and if $PO = U$, $p_1O = v_1$, and $p_2O = v_2$, we have for the first refraction

$$\frac{\mu' \cos^2 \phi'}{\mu_0 v_1} - \frac{\cos^2 \phi}{U} = \frac{\mu'}{\mu_0 v_2} - \frac{1}{U} = \frac{\sin(\phi - \phi')}{r_1 \sin \phi'}. \qquad (a)$$

For the second refraction ϕ' is the angle of incidence, ϕ is that of refraction, and $\frac{\mu_0}{\mu'}$ is the relative index, so if $q_1O = V_1$, and $q_2O = V_2$, we have

$$\frac{\mu_0 \cos^2 \phi}{\mu' V_1} - \frac{\cos^2 \phi'}{v_1} = \frac{\mu_0}{\mu' V_2} - \frac{1}{v_2} = \frac{\sin(\phi' - \phi)}{r_2 \sin \phi}$$

or
$$\frac{\cos^2 \phi}{V_1} - \frac{\mu' \cos^2 \phi'}{\mu_0 v_1} = \frac{1}{V_2} - \frac{\mu'}{\mu_0 v_2} = \frac{-\sin(\phi - \phi')}{r_2 \sin \phi'}$$

On adding (a) we obtain

$$\cos^2 \phi\left(\frac{1}{V_1} - \frac{1}{U}\right) = \frac{1}{V_2} - \frac{1}{U} = \frac{\sin(\phi - \phi')}{\sin \phi'}\left(\frac{1}{r_1} - \frac{1}{r_2}\right) \qquad (A)$$

This formula is, of course, only true for thin centric pencils; it is of no use when the whole aperture of a lens is employed to form an image, but only when a diaphragm with a small central perforation is used with the lens. It is, therefore, applicable to the case of a camera lens when a small "stop" is used to make the image more sharp, but perhaps the most important professional use of it is in the case of spectacles. The pupil of the eye then limits the width of the effective pencil, and when the visual lines traverse the centre of the spectacles an application of the formula gives rigidly accurate results. If the wearer gazes through eccentric portions of his glasses, the results are only approximately true, as then an investigation into the problem of oblique eccentric refraction would become necessary. However, for all practical purposes the simple formula we have obtained will suffice.

Short-sighted persons are often seen wearing their pince-nez tilted on their nose, and they declare that they see better with them in this position than when they are placed vertically. It will always be found in such cases that the myope is astigmatic, and that his pince-nez do not correct his astigmatism. He is, in fact, making use of this property of tilted lenses to obtain an astigmatic pencil which shall correct his error. I personally very frequently order tilted spectacles to poor patients to whom the expense of correct sphero-cylindrical lenses is prohibitive. Tilted lenses are of use when the power of the spherical part of the lens is high, and when only a weak cylindrical effect is required in vertical meridian, *i.e.* when the plane axis of the cylinder is horizontal; they have this incidental advantage, that they are lighter to wear.

We will take an example to illustrate this application of the formula, as it will show the most convenient way in which such questions may be treated.

Ex.—A patient wears his distance pince-nez (−10D) tilted 30° from the vertical plane. Calculate the effect of this displacement.

As the object is presumed to be at a great distance, the terms involving $\frac{1}{u}$ vanish, and remembering that

$$\frac{-D}{100} = \frac{1}{f''} = \overline{\mu - 1}\left(\frac{1}{r_1} - \frac{1}{r_2}\right)$$

when distances are measured in centimetres, we may write the equation (A) as

$$D_1 \cos^2 \phi = D_2 = \frac{D}{\mu - 1} \cdot \frac{\sin(\phi - \phi')}{\sin \phi'}$$

where D stands for −10 dioptres, D_1 for the dioptric power of the glass in the vertical meridian, and D_2 the dioptric power of the glass in the horizontal meridian.

When $\phi = 30°$ and $\mu = 1{\cdot}5$, D_2 is found to be very nearly −10·95 dioptres. Consequently,

$$D_1 = \frac{D_2}{\cos^2 \phi} = -\tfrac{4}{3}(10{\cdot}95) = -14{\cdot}6\text{D},$$

that is, the −10D glass with this tilt will act as if a cylindrical glass of power −3·65D, with the plane axis of the cylinder horizontal, had been added to a spherical lens of −10·95D. If we denote the cylindrical glass by D_c we may find its power more shortly thus

$$D_c = D_1 - D_2 = D_2\left(\frac{1 - \cos^2 \phi}{\cos^2 \phi}\right) = D_2 \tan^2 \phi$$

$$\therefore D_c = \tfrac{1}{3}(-10{\cdot}95) = -3{\cdot}65\text{D}.$$

Decentration of Lenses.—It is sometimes said that the refracting power of a lens is measured by the deviation that it induces on a ray of incident light. On referring to Fig. 59 this

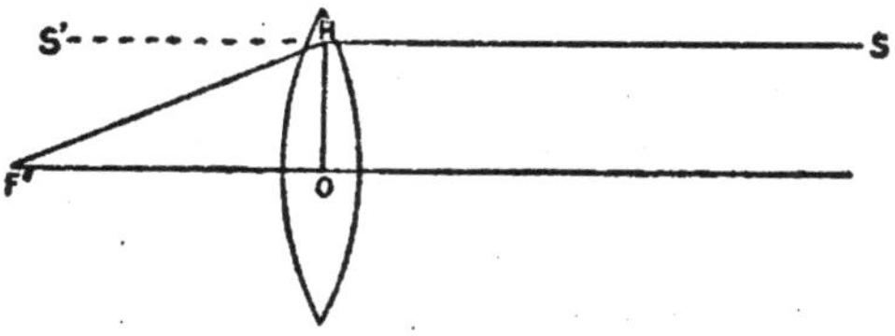

FIG. 59.

statement is seen to be untrue, or at any rate inadequate. The First Focus of the convex lens is indicated by F', and the ray F'H on emerging from the lens pursues the direction HS; its direction therefore undergoes a deviation F'HS' or HF'O, and it is clear that the deviation HF'O depends upon the height of OH, *i.e.* it depends upon which incident ray we choose to consider. It will be noticed that the angle HF'O is negative, being measured in the clockwise direction, so tan HF'O $= \frac{HO}{F'O}$ or $\frac{l}{f'}$ where HO is denoted by l. The lens, considered only with reference to the ray F'H has the same effect on it as a prism with its edge upwards; indeed, a lens may be regarded as a prism the strength of which is continually increasing as one passes from its centre O.

Now, if an eye were at S, it would see the image of an object at F' displaced to a great distance in the direction of S'. The lens would be said to be "decentred" HO millimetres downwards (with respect to the eye). If the lower part of the lens were cut away it would be called a Prismosphere, as it would act precisely like a spherical lens which had been bisected equatorially with a prism inserted between the two halves. Recently a great deal of attention has been directed to the properties and applications of decentred lenses or prismospheres, so that it may not be out of place to give the very simple formula which connects the decentration l of a lens with the deviation it induces.

Prism Dioptres, Centrads.—The Prism Dioptre is a unit proposed by Prentice, and is that angle whose tangent is 0·01 and is symbolized by delta (Δ). Consequently, with a +1D lens of which $f' = 1$ m. or 100 cm., when OH (Fig. 57) is 1 cm. a deviation of -1Δ is produced, for

$$\tan \text{HF'O} = \frac{-\text{OH}}{\text{F'O}} = \frac{-1}{100} = -1\Delta.$$

In the prescription of spectacles the decentration, HO or l, is always measured in millimetres, so that we have the simple formula $\text{N} = \frac{lD}{10}\Delta$, where N denotes the number of prism dioptres.

In order to obtain all the information possible from this formula we must make some special convention about signs. A prism, for instance, with its edge towards the right before the right eye would cause divergence of the eyes, but if so set before the left eye, it would cause convergence of the eyes. If, however, we agree to call prisms with their edges upwards or outwards (from the nose of the patient) negative, we shall get consistent results. Further, if we regard decentration downwards and inwards as also negative, the above simple formula will give us full and complete information about the prismatic equivalent of all decentred lenses. This convention about signs is a little difficult to remember; the *memoria technica* that I use is *De*centration, *d*own and *in*, *ne*gative; *P*risms *op*posite. For instance, what effect will a -5D lens have if decentred 4 mm. upwards? This means what effect will a spectacle lens have if its optical centre is displaced 4 mm. above the pupil of the eye?

Here $l = +4$, it is positive because the decentration is upwards, and

$$\text{N} = \frac{lD}{10}\Delta = \frac{4(-5)}{10}\Delta = -2\Delta$$

The decentration is therefore equivalent to a prism of 2 prism dioptres with its edge set in the negative direction, and from the *memoria technica* we know that must be either upwards or outwards, and from the conditions of the question we know that it must be upwards. The effect of this decentred lens will be the same as that of a normally centred -5D combined with a prism of 2Δ set edge upwards.

There are, however, some objections to the prism dioptre as a unit. It is not subject to the ordinary rules of arithmetic;

for instance, $2\Delta + 3\Delta$ are not exactly equivalent to 5Δ. This difficulty has been entirely obviated by Dennett's unit of the centrad, which is the hundredth part of a radian, and is denoted by a reversed delta (∇). Each unit is practically 34·377′, but the multiples of a centrad are appreciably greater than the multiples of a prism dioptre; and this is even of an advantage in using the formula $N = \frac{lD}{10}\nabla$, owing to spherical aberration. It has been adopted as the official unit by the American Ophthalmological Society, and before long I hope it will be universally used in this country. When the centrad is used, $3\nabla + 4\nabla$ are exactly equivalent to 7∇, and 10∇ is ten times the strength of 1∇.

What is the effect of a +2·5D lens decentred 4 mm. in?

$$N = \frac{lD}{10}\nabla = \frac{(-4)(2\cdot5)}{10}\nabla = -1\nabla.$$

The effect will be that of a normally centred +2·5D combined with a prism of one centrad set edge outwards (*i.e.* it will be abducting in function).

FORMULÆ FOR REFERENCE.

A. **Universally true—**

(*a*) For thin centric pencils—

(i) $\frac{f'}{p}+\frac{f''}{q}=1 \quad \therefore q=\frac{pf''}{p-f'}$ and $p=\frac{qf'}{q-f''}$.

(ii) $\frac{i}{o}=\frac{f'}{f'-p}=\frac{f''-q}{f''}$.

(*b*) For oblique pencils when v_1 and v_2 carry the same sign

$$v_2 \gtreqless v_1 \quad \text{as} \quad u \gtreqless \mu v_1$$

(*c*) Circle of least confusion

$$k=\frac{R(v_2-v_1)}{v_1+v_2-2s}$$

$$x=\frac{2v_1v_2-s(v_1+v_2)}{v_1+v_2-2s}$$

B. **Refraction at a Spherical Surface—**

When μ_0 is the index of the medium in which the source of light lies, and μ' is the index of the refracting medium—

$$\text{(i) } f'=\frac{-\mu_0 r}{\mu'-\mu_0}.$$

$$\text{(ii) } f''=\frac{\mu' r}{\mu'-\mu_0}=\frac{-\mu'}{\mu_0}f'.$$

Eccentric Pencils—

$$\frac{\mu'}{\mu_0}\cdot\frac{\cos^2\phi'}{v_1}-\frac{\cos^2\phi}{u}=\frac{\mu'}{\mu_0 v_2}-\frac{1}{u}=\frac{\sin(\phi-\phi')}{r\sin\phi'}.$$

C. **Lenses—**

(i) Thin, $\frac{1}{f''}=\frac{\mu'-\mu_0}{\mu_0}\left(\frac{1}{r_1}-\frac{1}{r_2}\right)=-\frac{1}{f'}$.

(ii) Thick, $\frac{1}{f''}=\frac{\mu'-\mu_0}{\mu_0 r_1 r_2}\left(r_2-r_1-\frac{\mu'-\mu_0}{\mu'}t\right)=-\frac{1}{f'}$.

Magnification—

$$M=\frac{\tan\theta'}{\tan\theta}:$$

(i) $M=\frac{l}{Kb}\left(1+\frac{q}{f'}\right)$. (ii) $M=\frac{l}{f'}$.

When $Kb=l=d+q$, (i) becomes $\frac{f'+q}{f'}$, (ii) becomes $\frac{d+q}{f'}$.

CARDINAL POINTS—

$$\text{When } K = t + f_1'' - f_2',$$

$$h' = \frac{f_1't}{K},\ h'' = \frac{f_2''t}{K},\ F' = \frac{-f_1'f_2'}{K},\ F'' = \frac{f_1''f_2''}{K}.$$

In complex systems, $t = \mathrm{H}_a''\mathrm{H}_b'$, $h' = \mathrm{H}'\mathrm{H}_a'$, $h'' = \mathrm{H}''\mathrm{H}_b''$.

DIOPTRES—

When f' given in centimetres $\mathrm{D} = \frac{100}{f'}$; when f' in millimetres,

$$D = \frac{1000}{f'}.$$

DECENTRATION AND CENTRADS—

$$\mathrm{N} = \frac{lD}{10}\nabla$$

THIN LENSES, OBLIQUE CENTRIC PENCILS—

$$\text{(i)}\ \cos^2\phi\left(\frac{1}{\mathrm{V}_1} - \frac{1}{\mathrm{U}}\right) = \frac{1}{\mathrm{V}_2} - \frac{1}{\mathrm{U}} = \frac{\sin(\phi - \phi')}{\sin\phi'}\left(\frac{1}{r_1} - \frac{1}{r_2}\right).$$

(ii) When incident rays are parallel—

$$D_2 = \frac{D}{\mu - 1}\cdot\frac{\sin(\phi - \phi')}{\sin\phi'},\qquad D_c = D_2\tan^2\phi.$$

D. Prisms—

$$\mathrm{D} = \psi - \phi - \mathrm{A}.$$

MINIMUM DEVIATION—

$$\mathrm{D} = 2\psi - \mathrm{A}.$$

$$\mu = \frac{\sin\frac{1}{2}(\mathrm{A} + \mathrm{D})}{\sin\frac{1}{2}\mathrm{A}}.$$

If A small, $\mathrm{D} = \overline{\mu - 1}\,\mathrm{A}.$

In order to convert the formulæ (B) into those suitable for a plane surface, put $r = \infty$. To convert them into the corresponding formulæ for reflection, put $\mu = -1$ and $\phi' = -\phi$.

ANSWERS

CHAPTER I (p. 10).

(1) 45 feet. (2) 70 feet.

(3) The intensity of the electric light is 900 times that of the gaslight.

(4) The shadow cast by the gas-flame will be approximately 8 times more intense than that cast by the electric light.

CHAPTER II (p. 17).

(1) 1 foot, 5 feet, 7 feet. (2) 7 images.

(3) One image. If ECB $< 60°$, two images would be seen. For PCB $= 120°$ since ACP $= 15°$ and ACB $= 135°$. Produce BC to c, and draw Pp^1 perpendicular to BC produced, cutting it at c; make $cp^1 = Pc$; then the angle $p^1Cc = 60°$. Consequently, the line p^1E will only cut the mirror when ECB $< 60°$.

CHAPTER III (p. 33).

(1) The image is 15 cm. in height, it is virtual and erect, and it is formed $16\frac{2}{3}$ cm. behind the mirror.

(2) (i.) 24 inches in front of the mirror.

(ii.) 8 inches behind the mirror.

(3) $r = -6$ inches; the mirror is convex.

(4) $p = 6\frac{6}{11}$ inches, $\frac{i}{o} = -11$; the image is inverted.

CHAPTER IV (p. 48).

(2) $\mu = 1{\cdot}54$. (3) $\mu = \sqrt{2}$.

(5) Two prisms of minimum deviations 1° 44′ and 2° respectively, if properly placed, would correct the deviation. The weaker prism should be placed edge outwards before the left eye, while the 2° prism should be placed edge outwards and upwards before the right eye, so that the base apex line makes with the horizontal line an angle of 30°.

CHAPTER V (p. 67).

(1) An inverted image $-\frac{2}{3}$ mm. high would be formed $-22{\cdot}2$ mm. from the cornea, *i.e.* $2{\cdot}\dot{2}$ mm. behind the second principal focus.

(2) $\dfrac{f'}{p}+\dfrac{f''}{q}=1$ or $\dfrac{f'}{p}-\dfrac{\mu f'}{q}=1, \therefore \dfrac{f'}{150}+\dfrac{f'}{15}=1, \therefore f'=\dfrac{150}{11}$.

$r=-(\mu-1)f'=-\frac{50}{11}$.

$\dfrac{i}{o}=\dfrac{f'}{f'-p}=\dfrac{150}{-1500}=-\dfrac{1}{10}, \therefore I_c=-0{\cdot}6$

$\dfrac{I_a}{I_c}=\dfrac{-0{\cdot}\dot{6}}{-0{\cdot}6}=\dfrac{10}{9}$.

(3) A real image will be formed at the distance of the diameter of the sphere from the unsilvered side.

CHAPTER VI (p. 107).

(1) $\dfrac{i}{o}=-\frac{1}{3}, q=-7{\cdot}2$ inches.

(2) $\dfrac{i}{o}=\dfrac{f'}{f'-p}=\dfrac{f''-q}{f''}, \therefore f'f''=(f'-p)(f''-q)$.

(3) $r_2=39$ inches; in water $f'=37\frac{1}{2}$ inches.

(4) (i) $M=48{\cdot}5$, (ii) $M=48{\cdot}5$, (iii) $M=50$.

(5) h' or $H'A=-1{\cdot}6$, h'' or $H''D=0{\cdot}8$, $F'H'=3{\cdot}2$, $F''H''=-3{\cdot}2$.

INDEX

THE END

PRINTED BY WILLIAM CLOWES AND SONS, LIMITED, LONDON AND BECCLES.

Spotted something that isn't quite right?

Let us know and we'll fix it for future reprints. And as a thank you for your diligence, we'll send you a free Sothis book of your choice.

publisher@sothis.press

This page intentionally left blank

Made in the USA
Coppell, TX
04 November 2026

11203176R20083